SpringerBriefs in Mathematics

T0203118

SpringerBriefs in Mathematics showcases expositions in all areas of mathematics and applied mathematics. Manuscripts presenting new results or a single new result in a classical field, new field, or an emerging topic, applications, or bridges between new results and already published works, are encouraged. The series is intended for mathematicians and applied mathematicians.

For further volumes:
http://www.springer.com/series/10030

Alain Bensoussan • Jens Frehse • Phillip Yam

Mean Field Games and Mean Field Type Control Theory

 Springer

Alain Bensoussan
Naveen Jindal School of Management
University of Texas at Dallas
Richardson, TX, USA

Department of Systems Engineering
    and Engineering Management
City University of Hong Kong
Kowloon, Hong Kong SAR

Phillip Yam
Department of Statistics
The Chinese University of Hong Kong
Shatin, Hong Kong SAR

Jens Frehse
Institut für Angewandte Mathematik
Universitat Bonn
Bonn, Germany

ISSN 2191-8198   ISSN 2191-8201 (electronic)
ISBN 978-1-4614-8507-0   ISBN 978-1-4614-8508-7 (eBook)
DOI 10.1007/978-1-4614-8508-7
Springer New York Heidelberg Dordrecht London

Library of Congress Control Number: 2013945868

Mathematics Subject Classification (2010): 49J20, 49N90, 58E25, 91G80, 35Q91, 49N70, 49N90, 91A06, 91A13, 91A18, 91A25, 35R15, 60H30, 35R60, 60H15, 60H30, 91A15, 93E20

Printed on acid-free paper

Springer is part of Springer Science+Business Media (www.springer.com)

Preface

Mean field theory has raised a lot of interest in recent years, since the independent introduction by Lasry–Lions and Huang–Caines–Malhamé, see, in particular, Lasry–Lions [25–27], Gueant et al. [17], Huang et al. [21,22], Buckdahn et al. [13], Andersson–Djehiche [1], Cardaliaguet [14], Carmona–Delarue [15], Bensoussan et al. [10]. The applications concern approximating an infinite number of players with common behavior by a representative agent. This agent has to solve a control problem perturbed by a field equation, representing in some way the average behavior of the infinite number of agents. The mean field term can influence the dynamics of the state equation of the agent as well as his/her objective functional. In the mean field game, the agent cannot influence the mean field term, considered as external. Therefore, he or she solves a standard control problem, in which the mean field term acts as a parameter. In this context one looks for an equilibrium, which means that the mean field term is the probability distribution of the state behavior of the individual agent. The equilibrium is the core of the mathematical difficulty.

In the mean field type control problem, the agent can influence the mean field term. The problem is thus a control problem, albeit more elaborate than the standard control theory problem. Indeed, the state equation also contains the probability distribution of the state and thus is of the McKean–Vlasov type; see McKean [29].

The objective of this book is to describe the major approaches to two types of problems and more advanced questions for future research. In this framework, we are not presenting full proofs of results, but we describe what are the mathematical problems, where they come from, and the steps to be accomplished to obtain solutions.

Richardson, TX, USA
Bonn, Germany
Shatin, Hong Kong SAR

Alain Bensoussan
Jens Frehse
Phillip Yam

Acknowledgments

Alain Bensoussan expresses his gratitude to the financial supports from the National Science Foundation under grant DMS-1303775, and Research Grant Council of Hong Kong Special Administrative Region under grant GRF 500113. Phillip Yam also acknowledges the financial supports from The Hong Kong RGC GRF 404012 with the project title "Advance Topics in Multivariate Risk Management in Finance and Insurance", and The Chinese University of Hong Kong Direct Grant 2011/2012 Project ID 2060444. Phillip Yam also expresses his sincere gratitude to the hospitality of Hausdorff Center of Mathematics of University of Bonn for his fruitful stay in Hausdorff Trimester Program with title: "Stochastic Dynamics in Economics and Finance".

Contents

Chapter 1
Introduction

Mean field games and mean field type control introduce new problems in control theory. The term "games" is appealing, although maybe confusing. In fact, mean field games are control problems, in the sense that one is interested in a single decision maker, who we call the representative agent. However, these problems are not standard, since both the evolution of the state and the objective functional are influenced by terms that are not directly related to the state or to the control of the decision maker. They are, however, indirectly related to the decision maker, in the sense that they model a very large community of agents similar to the representative agent. All the agents behave similarly and impact the representative agent. However, because of the large number, an aggregation effect takes place. The interesting consequence is that the impact of the community can be modelled by a mean field term, but when this is done the problem is reduced to a control problem. Note that the concept of mean field is very fruitful in physics. However, applying the idea of averaging to domains different from physics is the novelty. The idea is also different from the concept of equilibrium in economics because of both the emphasis on dynamic aspects (as in physics) and the direct effect of the mean field term on the state evolution of the agent.

Of course, an important question is whether the mean field control problem is a good approximation, and if so, of what exactly. This is an essential aspect of the theory. It can be done for mean field games and explains the terminology. What can be expected is that if one considers a Nash game for identical agents, the equilibrium can be approximated by the solution of the mean field control equilibrium, whenever all agents use the same control, defined by the optimal control of the representative agent. The approximation improves as the number of agents increases. This basic result is the justification of the concept of mean field games. The interest of the mean field type control problem is that it is a control problem and not an equilibrium. Consequently, a solution may be found more often (at least some approximate solution).

So to sum up, mean field games can be reduced to a standard control problem and an equilibrium, and mean field type control is a nonstandard control problem.

A. Bensoussan et al., *Mean Field Games and Mean Field Type Control Theory*,
SpringerBriefs in Mathematics, DOI 10.1007/978-1-4614-8508-7_1,
© Alain Bensoussan, Jens Frehse, Phillip Yam 2013

To solve these problems, one naturally can rely on the techniques of control theory: dynamic programming and the stochastic maximum principle. It is interesting to observe that in the literature those papers dealing with mean field games use dynamic programming and those that deal with mean field type control use the stochastic maximum principle. The reason is that in the mean field games problem, the control problem is standard, since the mean field term is external. It is then natural to write a Hamilton–Jacobi–Bellman (HJB) equation. The optimal control is derived from a feedback. The mean field terms involve a probability distribution, which in the equilibrium is governed by a Fokker–Planck (FP) equation. It represents the probability distribution of the state. This equation depends on the feedback and also on the solution of the HJB equation. So the problem becomes a coupled system of partial differential equations, one HJB equation, and one Fokker–Planck equation. This is the mathematical problem to be solved. Of course, for a standard control problem, one can use the stochastic maximum principle as an alternative to the HJB equation. This has been done recently in [15] and [10]. One obtains again a coupling, and there is a fixed point to be sought. In the mean field type control problem, one uses the stochastic maximum principle and not dynamic programming, because the control problem is not standard. In particular, there is "time inconsistency". The optimality principle of dynamic programming is not valid, and therefore one cannot write an HJB equation for such a nonstandard stochastic control problem. Since the stochastic maximum principle does not rely on the dynamic programming optimality principle, the time inconsistency is not an issue.

We show in this book that we can obtain an HJB equation, but only when coupled with a Fokker-Planck equation. This is not a contradiction of the previous discussion. What cannot be obtained is a single HJB equation, because the optimality principle is not valid. However a coupled system of HJB and Fokker–Planck equations can be obtained, exactly like those in the case of mean field games. We show that this result is fully consistent with the results obtained from the stochastic maximum principle, and in fact the stochastic maximum principle can be recovered by this approach.

So this is the first contribution of this book—to present a unified approach for both problems, using either HJB- and FP-coupled equations or the stochastic maximum principle.

It is important to stress that the coupled system of partial differential equations (HJB-FP) is different for both problems. The HJB-FP system for the mean field type control problem contains additional terms, so it is more complex than that of the mean field game. It had not appeared earlier in the literature, and therefore no results were known. However, the general structures of the two coupled systems are similar, so one can expect solutions for both, even paradoxically more often for the mean field type problem, because it is a control problem and not an equilibrium. We shall show this in the case of linear quadratic problems, for which everything is explicit.

We then explain in the case of mean field games that what is to be done is to show that the optimal solution derived from the coupled HJB-FP equations can be used as a good approximation for the Nash equilibrium of a large number of identical agents. Recalling that the representative agent gets an optimal feedback, the most natural way to design the Nash equilibrium is to use identical feedbacks for all agents. This works provided that the feedback is sufficiently smooth, and thus one obtains a Nash equilibrium among smooth feedbacks. To avoid this smoothness difficulty, one can introduce open loop controls derived from the feedback, and then consider a Nash equilibrium among these open loop controls.

We conjecture that the mean field type optimal control can also be used to provide an approximation for the Nash equilibrium of a large number of agents. However, no such result exists in the literature.

This issue of "time inconsistency" has been considered in the literature independently of the mean field theory, see Björk-Murgoci, [11], Hu-Jin-Zhou [18]. The objective is to consider feedbacks that do not depend on the initial condition, such as in standard control theory. A solution that depends on the initial conditions is labelled "precommittment" in the economics literature. One way to obtain only time-consistent feedbacks is to reformulate the control problem into a game problem. One considers decisions made at future times as done by different players, one player at a time. Therefore, at each time, the decision maker is a player who looks for a Nash equilibrium against players at future times. We shall see that one can handle only a limited number of situations, as far as coupling is concerned. Another way to understand the concept is to consider optimality against "spike modifications."

We next review the situation of linear quadratic problems. The interesting aspect is that all solutions can be obtained explicitly and thus can be compared. It is particularly useful to compare the assumptions. They are not at all equivalent. We do not have the situation in which we can identify the best approach. It is a case-by-case situation.

We explain next why the coupled HJB-FP system is written naturally as a parabolic system, even when we consider an infinite horizon.

To derive stationary (elliptic) systems, Lasry-Lions have considered an ergodic control case. We show in this book that it is possible to consider elliptic coupled HJB-FP equations, without using ergodic situations. The problems are less natural but we gain the simplicity.

In fact, it is then possible to benefit in this framework from other interpretations of the coupled HJB-FP system. If one considers the dual control problem, then the state equation is the Fokker–Planck equation, describing the probability distribution of the state, while the HJB equation can be interpreted as a necessary condition of optimality for the dual problem. To generate mean field terms, it is sufficient to consider objective functions that are not just linear in the probability distribution, but are also more complex.

This approach also has a different type of application. In the traditional stochastic control problem, the objective functional is the expected value of a cost depending on the trajectory. So it is linear in the probability measure. This type of functional leaves out many current considerations in control theory, namely situations where one wants to take into consideration not just the expected value but also the variance. This case occurs often in risk management.

The famous mean-variance optimization problem (i.e., the Markowitz problem) is an example. In addition, one may be interested in several functionals along the trajectory, even though one may be satisfied with expected values. Combining these various expected values into a single payoff, one is led naturally to mean field problems. They are meaningful even without considering ergodic theory, i.e., long-term behavior.

We then address future important extensions.

In most real problems of economics, there is not just one representative agent and a large community of identical players, which bring impact via a mean field term. There are several major/dominating players, as well as large communities.

So a natural question is to consider the problem of these major players. They know that they can influence the community, and they also compete with each other. So the issue is that of differential games, with mean field terms, and not of mean field equations arising from the limit of a Nash equilibrium for an infinite number of players. Huang–Caines–Malhamé [21] allow for groups of players with different characteristics, but they do not compete with one another. Huang [19] and Nourian-Caines [31] study the situation of a major player. The representative agent is submitted to the major player. The major player will take this into account in his or her decision. The problem has similarities with Stackelberg games. However, the state probabilities have to be replaced with conditional probabilities, which is much more complex.

In the context of coalitions competing with each other, the objective in this work is to present systems of HJB equations, coupled with systems of FP equations. We explain how they can be obtained from averaging large homogeneous communities, who compete with one another. This type of problem has not yet been addressed in the literature, except in the paper by two of the authors, [7].

To recover the system of nonlinear PDEs it is easier to proceed with the dual problems as explained above. One can consider a differential game for state equations that are probability distributions of states and evolve according to FP equations. One recovers nonlinear systems of PDEs with mean field terms, with a different motivation. An additional interesting feature of this approach is that we do not need to consider an ergodic situation. In fact,considering strictly positive discounts is quite meaningful in our applications. This leads to systems of nonlinear PDEs with mean field coupling terms, which we can study with a minimum set of assumptions. The ergodic case, when the discount vanishes, requires much more stringent assumptions, as is already the case when there is no mean field term. We refer to Bensoussan–Frehse [5, 6] and Bensoussan–Frehse–Vogelgesang [8, 9] for the situation without the mean field terms. Basically our set of assumptions remains valid and we have to incorporate additional assumptions to deal with the mean field terms.

We then provide an overview of the analytic techniques to solve for the systems of HJB-FP equations. We do it only in a limited number of situations.

An interesting aspect of our approach is to proceed with a priori estimates. We use fixed-point theory only for approximations, which is much easier.

It is clear that a lot remains to be done in developing models, techniques, and applications. For instance, we have not considered the issue of systemic risk; see [16], which bears similarity with mean field theory. The objective is, however, different. In systemic risk, one is interested in the consequences of a random shock on a equilibrium within a network of interactions. Similarly, among the interesting new techniques, let us mention sensitivity, see [23,24], which leads to useful results. Unfortunately, within the page limitation of this current discussion, we could not discuss these possible applications.

We hope that, in spite of its limitations, the present synthesis will further understanding of the diversity of concepts and problems. We want to express our gratitude to those who initiated the domain, Caines–Huang–Malhamé on the one hand, and Lasry–Lions on the other hand, for their inspiring work. The number of papers that have originated from their initial articles and lectures is the best evidence of the importance of their ideas.

Chapter 2
General Presentation of Mean Field Control Problems

2.1 Model and Assumptions

Consider a probability space $(\Omega, \mathscr{A}, P)$ and a filtration $\mathscr{F}^t$ generated by a n-dimensional standard Wiener process $w(t)$. The state space is $\mathbb{R}^n$ with the generic notation x and the control space is $\mathbb{R}^d$ with generic notation v. We consider measurable functions

$$g(x,m,v) : \mathbb{R}^n \times L^1(\mathbb{R}^n) \times \mathbb{R}^d \to \mathbb{R}^n; \quad \sigma(x) : \mathbb{R}^n \to \mathscr{L}(\mathbb{R}^n; \mathbb{R}^n);$$

$$f(x,m,v) : \mathbb{R}^n \times L^1(\mathbb{R}^n) \times \mathbb{R}^d \to \mathbb{R}; \quad h(x,m) : \mathbb{R}^n \times L^1(\mathbb{R}^n) \to \mathbb{R}. \tag{2.1}$$

These functions may depend on time. We omit this dependence to save notation. We assume that

$$\sigma(x), \sigma^{-1}(x) \text{ are bounded.} \tag{2.2}$$

The argument m will accommodate the mean field term. In practice, it will be a probability density on $\mathbb{R}^n$. One could replace $L^1(\mathbb{R}^n)$ by a space $L^p(\mathbb{R}^n)$, $1 < p < \infty$. At some point, we shall need to extend these functions to arguments that are probability measures, not having densities with respect to the Lebesgue measure; in practice, for example, a finite sum of Dirac measures. The space of probability measures will be equipped with the topology of weak $*$ convergence. In this situation we lose the structure of Banach space. So we stick to densities as much as possible. Consider a function $m(t) \in C(0, T; L^1(\mathbb{R}^n))$. We pick a feedback control in the form $v(x,t)$, a measurable map from $\mathbb{R}^n \times (0, T)$ to $\mathbb{R}^d$. To save notation we shall write it $v(x)$. We solve the stochastic differential equation (SDE)

$$dx = g(x(t), m(t), v(x(t)))dt + \sigma(x(t))dw(t),$$

$$x(0) = x_0. \tag{2.3}$$

A. Bensoussan et al., *Mean Field Games and Mean Field Type Control Theory*, SpringerBriefs in Mathematics, DOI 10.1007/978-1-4614-8508-7_2, © Alain Bensoussan, Jens Frehse, Phillip Yam 2013

We will need to assume that

$$g, \sigma, f, h \quad \text{are differentiable in both } x, v. \tag{2.4}$$

With this assumption, if the feedback is Lipshitz, the SDE has a unique solution $x(t)$, which is a continuous process adapted to the filtration $\mathscr{F}^t$. This is the state of the system under the control defined by the feedback $v(.)$. The initial state x_0 is a random variable independent of the Wiener process, which has a probability density is given by m_0. Note that the function $m(t)$ acts as a parameter. We assume, however, that $m(0) = m_0$.

To the pair $(v(.), m(.))$ we associate the control objective

$$J(v(.), m(.)) = E\left[\int_0^T f(x(t), m(t), v(x(t)))\, dt + h(x(T), m(T))\right]. \tag{2.5}$$

To define this functional properly, we need to assume that

$$g \text{ has linear growth in } x, v.$$

$$f, h \text{ have quadratic growth in } x, v. \tag{2.6}$$

Recall that $m(t)$ is a deterministic function.

2.2 Definition of the Problems

The mean field game problem is defined as follows: find a pair $(\hat{v}(.), m(.))$ such that, denoting by $\hat{x}(.)$ the solution of

$$d\hat{x} = g(\hat{x}(t), m(t), \hat{v}(\hat{x}(t)))dt + \sigma(\hat{x}(t))dw(t),$$
$$\hat{x}(0) = x_0, \tag{2.7}$$

then

$$m(t) \text{ is the probability distribution of } \hat{x}(t), \forall t \in [0, T]$$
$$J(\hat{v}(.), m(.)) \leq J(v(.), m(.)) \forall v(.). \tag{2.8}$$

The mean field type control problem is defined as follows: For any feedback $v(.)$, let $x(t) = x_{v(.)}(t)$ be the solution of (2.3) with $m(t) = $ probability distribution of $x_{v(.)}(t)$. So (2.3) becomes a McKean–Vlasov equation. If we denote by $m_{v(.)}(t) = $ probability distribution of $x_{v(.)}(t)$, we thus have

$$dx_{v(.)} = g(x_{v(.)}(t), m_{v(.)}(t), v(x_{v(.)}(t))dt + \sigma(x_{v(.)}(t))dw(t),$$
$$x(0) = x_0. \tag{2.9}$$

$$m_{v(.)}(t) = \text{probability distribution of } x_{v(.)}(t) \tag{2.10}$$

then we have to find $\hat{v}(.)$ such that

$$J(\hat{v}(.), m_{\hat{v}(.)}(.)) \leq J(v(.), m_{v(.)}(.)) \ \forall v(.). \tag{2.11}$$

If we denote $\hat{x}(t) = x_{\hat{v}(.)}(t)$ and $m(t) = m_{\hat{v}(.)}(t)$, then we can write

$$m(t) \text{ is the probability distribution of } \hat{x}(t), \forall t \in [0, T]$$

$$J(\hat{v}(.), m(.)) \leq J(v(.), m_{v(.)}(.)), \forall v(.). \tag{2.12}$$

It is useful to compare (2.8) with (2.12). The difference takes place in the right-hand side of the inequality in the second condition. For a mean field game, $m(.)$ is fixed, whereas for mean field type control, m depends on $v(.)$.

Remark 1. We can see that in our control problems we do not control the term $\sigma(x)$. There is not either the term m in it. This will considerably simplify the mathematical treatment. Besides, there is no degeneracy, This will allow us to work with densities, and to benefit from the nice structure of $L^1(\mathbb{R}^n)$ instead of having to work with the space of probability measures on $\mathbb{R}^n$.

Chapter 3
The Mean Field Games

3.1 HJB-FP Approach

Let us set

$$a(x) = \frac{1}{2}\sigma(x)\sigma^*(x),$$ (3.1)

and introduce the second-order differential operator

$$A\varphi(x) = -\mathrm{tr}\; a(x)D^2\varphi(x).$$ (3.2)

We define the dual operator

$$A^*\varphi(x) = -\sum_{k,l=1}^{n} \frac{\partial^2}{\partial x_k \partial x_l}(a_{kl}(x)\varphi(x)).$$ (3.3)

Since $m(t)$ is the probability distribution of $\hat{x}(t)$, it has a density with respect to the Lebesgue measure denoted by $m(x,t)$, which is the solution of the Fokker–Planck equation

$$\frac{\partial m}{\partial t} + A^*m + \mathrm{div}\,(g(x,m,\hat{v}(x))m) = 0,$$

$$m(x,0) = m_0(x).$$ (3.4)

We next want the feedback $\hat{v}(x)$ to solve a standard control problem, in which m appears as a parameter. We can thus readily associate an HJB equation with this problem, parametrized by m. We introduce the Lagrangian function

$$L(x,m,v,q) = f(x,m,v) + q \cdot g(x,m,v),$$ (3.5)

and the Hamiltonian function

$$H(x,m,q) = \inf_v L(x,m,v,q).$$ (3.6)

A. Bensoussan et al., *Mean Field Games and Mean Field Type Control Theory*, SpringerBriefs in Mathematics, DOI 10.1007/978-1-4614-8508-7_3,

11

In this context, we shall assume that the infimum is attained and that we can define a sufficiently smooth function $\hat{v}(x, m, q)$ such that

$$H(x, m, q) = L(x, m, \hat{v}(x, m, q), q). \tag{3.7}$$

We shall also write

$$G(x, m, q) = g(x, m, \hat{v}(x, m, q)). \tag{3.8}$$

From standard control theory, the HJB equation is defined by

$$-\frac{\partial u}{\partial t} + Au = H(x, m, Du),$$

$$u(x, T) = h(x, m(T)), \tag{3.9}$$

and setting $\hat{v}(x) = \hat{v}(x, m, Du)$, we can assert, from standard arguments,

$$J(\hat{v}(.), m(.)) = \int_{\mathbb{R}^n} u(x, 0) m_0(x) dx \leq J(v(.), m(.)), \ \forall v(.) \tag{3.10}$$

Therefore, to solve the mean field game problem, we have to solve the coupled HJB-FP system of PDEs

$$-\frac{\partial u}{\partial t} + Au = H(x, m, Du)$$

$$\frac{\partial m}{\partial t} + A^* m + \text{div} \ (G(x, m, Du)m) = 0$$

$$u(x, T) = h(x, m(T))$$

$$m(x, 0) = m_0(x) \tag{3.11}$$

and the pair $\hat{v}(x) = \hat{v}(x, m, Du)$, m is the solution. We can then proceed backwards as usual in control theory. We solve the system (3.11), which gives us a candidate $\hat{v}, m$, and then rely on a verification argument to check that it is indeed the solution.

Since we rely on PDE techniques, it is essential that the operator A be linear and not degenerate.

3.2   Stochastic Maximum Principle

We use notation that is customary in stating the stochastic maximum principle. From the feedback $\hat{v}(.)$ and the probability distribution $m(t)$, we construct stochastic processes $X(t) \in \mathbb{R}^n$, $V(t) \in \mathbb{R}^d$, $Y(t) \in \mathbb{R}^n$, $Z(t) \in \mathscr{L}(\mathbb{R}^n; \mathbb{R}^n)$ which are defined as follows

$$X(t) = \hat{x}(t), \ m(t) = P_{X(t)}$$

in which the notation $P_{X(t)}$ means the probability distribution of the random variable $X(t)$. We next define

$$Y(t) = Du(X(t),t), \; V(t) = \hat{v}(X(t),P_{X(t)},Y(t))$$

and finally

$$Z(t) = D^2 u \, \sigma(X(t),t).$$

We first have from (3.1)

$$dX = g(X(t),P_{X(t)},V(t))dt + \sigma(X(t))dw(t)$$

$$X(0) = x_0. \tag{3.12}$$

Using Ito's formula, we next write

$$dY_i(t) = \sum_k \frac{\partial^2 u}{\partial x_i \partial x_k} dX_k + \left(\frac{\partial^2 u}{\partial x_i \partial t} + \sum_{kl} a_{kl} \frac{\partial^3 u}{\partial x_i \partial x_k \partial x_l} \right) dt.$$

We differentiate in x_i the HJB equation to evaluate $\dfrac{\partial^2 u}{\partial x_i \partial t}$. We get

$$-\frac{\partial^2 u}{\partial x_i \partial t} - \sum_{kl} \frac{\partial a_{kl}}{\partial x_i} \frac{\partial^2 u}{\partial x_k \partial x_l} - \sum_{kl} a_{kl} \frac{\partial^3 u}{\partial x_i \partial x_k \partial x_l}$$

$$= \frac{\partial f}{\partial x_i}(x,m,\hat{v}(x)) + \sum_k g_k(x,m,\hat{v}(x)) \frac{\partial^2 u}{\partial x_i \partial x_k} + \sum_k \frac{\partial u}{\partial x_k} \frac{\partial g_{x_k}}{\partial x_i}(x,m,\hat{v}(x)).$$

Combining terms and using the definition of $Y(t),Z(t)$ we obtain easily

$$-dY = \left(\frac{\partial f}{\partial x}(X(t),P_{X(t)},V(t)) + \frac{\partial g}{\partial x}^*(X(t),P_{X(t)},V(t))Y(t) \right.$$

$$\left. + \operatorname{tr} \frac{\partial \sigma(X(t))}{\partial x}^* Z(t) \right) dt - Z(t)dw(t),$$

$$Y(T) = \frac{\partial h(X(T),P_{X(T)})}{\partial x}.$$

In the context of the stochastic maximum principle, it is customary to call Hamiltonian what is called Lagrangian in the context of dynamic programming. Because of this, we call Hamiltonian

$$H(x,m,v,q) = f(x,m,v) + q.g(x,m,v). \tag{3.13}$$

We can collect results and summarize the stochastic maximum principle as follows: find adapted processes

$$X(t) \in R^n, V(t) \in R^d, Y(t) \in R^n, Z(t) \in \mathscr{L}(R^n; R^n)$$

$$dX = g(X(t), P_{X(t)}, V(t))dt + \sigma(X(t))dw(t),$$

$$-dY = \left(\frac{\partial H}{\partial x}(X(t), P_{X(t)}, V(t), Y(t)) + \operatorname{tr} \frac{\partial \sigma(X(t))^*}{\partial x} Z(t) \right) dt - Z(t)dw(t),$$

$$X(0) = x_0, \quad Y(T) = \frac{\partial h(X(T), P_{X(T)})}{\partial x}. \tag{3.14}$$

$$V(t) \text{ minimizes } H(X(t), P_{X(t)}, v, Y(t)) \text{ in } v. \tag{3.15}$$

This is a forward–backward SDE of the McKean–Vlasov type.

The usual approach is to not derive this problem from the HJB-FP system. This requires a lot of smoothness, which is not really necessary. The usual approach is to study (3.14), (3.16) directly, by probabilistic techniques (see [15, 31]).

Example 2. Suppose

$$f(x,m,v) = f(x, \int_{R^n} \varphi(\xi)m(\xi)d\xi, v), \quad h(x,m) = h\left(x, \int_{R^n} \psi(\xi)m(\xi)d\xi \right)$$

$$g(x,m,v) = g(x, \int_{R^n} \chi(\xi)m(\xi)d\xi, v), \tag{3.16}$$

where φ, ψ, χ map R^n into R^n. By abuse of notation, we can describe in the same way the functions $f(x, \xi, v)$ on $\mathbb{R}^n \times \mathbb{R}^n \times \mathbb{R}^d$ and $f(x, m, v)$ on $\mathbb{R}^n \times L^1(\mathbb{R}^n) \times \mathbb{R}^d$, and similarly for h and g. We can then write the stochastic maximum principle (3.15) as follows

$$X(t) \in \mathbb{R}^n, V(t) \in \mathbb{R}^d, Y(t) \in \mathbb{R}^n, Z(t) \in \mathscr{L}(\mathbb{R}^n; \mathbb{R}^n)$$

$$dX = g(X(t), E\chi(X(t)), V(t))dt + \sigma(X(t))dw(t),$$

$$-dY = \left(\frac{\partial f}{\partial x}(X(t), E\varphi(X(t)), V(t)) + \frac{\partial g^*}{\partial x}(X(t), E\chi(X(t)), V(t))Y(t) \right.$$

$$\left. + \operatorname{tr} \frac{\partial \sigma(X(t))^*}{\partial x} Z(t) \right) dt - Z(t)dw(t),$$

$$X(0) = x_0, \quad Y(T) = \frac{\partial h(X(T), E\psi(X(T)))}{\partial x}. \tag{3.17}$$

$$V(t) \text{ minimizes } f(X(t), E\varphi(X(t)), v) + Y(t).g(X(t), E\chi(X(t)), v) \text{ in } v. \tag{3.18}$$

Chapter 4
The Mean Field Type Control Problems

4.1 HJB-FP Approach

We need to assume that the

$$m \to f(x,m,v), \; g(x,m,v), \; h(x,m)$$

$$\text{are differentiable in } m \in L^2(\mathbb{R}^n) \tag{4.1}$$

and we use the notation $\dfrac{\partial f}{\partial m}(x,m,v)(\xi)$ to represent the derivative, so that

$$\frac{d}{d\theta} f(x, m + \theta \tilde{m}, v)_{|\theta=0} = \int_{\mathbb{R}^n} \frac{\partial f}{\partial m}(x,m,v)(\xi) \tilde{m}(\xi) d\xi.$$

Here, x, v are simply parameters. Coming back to the definition (2.9)–(2.11), consider a feedback $v(x)$ and the corresponding trajectory defined by (2.9). The probability distribution $m_{v(.)}(t)$ of $x_{v(.)}(t)$ is a solution of the FP equation

$$\frac{\partial m_{v(.)}}{\partial t} + A^* m_{v(.)} + \operatorname{div}\,(g(x, m_{v(.)}, v(x)) m_{v(.)}) = 0,$$

$$m_{v(.)}(x,0) = m_0(x) \tag{4.2}$$

and the objective functional $J(v(.), m_{v(.)})$ can be expressed as follows

$$J(v(.), m_{v(.)}(.)) = \int_0^T \int_{\mathbb{R}^n} f(x, m_{v(.)}(x), v(x)) m_{v(.)}(x) dx dt$$

$$+ \int_{\mathbb{R}^n} h(x, m_{v(.)}(x,T)) m_{v(.)}(x,T) dx. \tag{4.3}$$

A. Bensoussan et al., *Mean Field Games and Mean Field Type Control Theory*, SpringerBriefs in Mathematics, DOI 10.1007/978-1-4614-8508-7_4, © Alain Bensoussan, Jens Frehse, Phillip Yam 2013

Consider an optimal feedback $\hat{v}(x)$ and the corresponding probability density $m_{\hat{v}(.)}(x,t) = m(x,t)$. Then let $v(.)$ be any feedback and $\hat{v}(x) + \theta v(x)$. We want to compute

$$\frac{dm_{\hat{v}(.)+\theta v(.)}(x)}{d\theta}\bigg|_{\theta=0} = \tilde{m}(x).$$

We can check that

$$\frac{\partial \tilde{m}}{\partial t} + A^*\tilde{m} + \text{div}\,(g(x,m,\hat{v}(x))\tilde{m})$$

$$+ \text{div}\left(\left[\int \frac{\partial g}{\partial m}(x,m,\hat{v}(x))(\xi)\tilde{m}(\xi)d\xi + \frac{\partial g}{\partial v}(x,m,\hat{v}(x))v(x)\right]m(x)\right) = 0,$$

$$\tilde{m}(x,0) = 0. \tag{4.4}$$

It then follows that

$$\frac{dJ(\hat{v}(.)+\theta v(.), m_{\hat{v}(x)+\theta v(x)}(.))}{d\theta}\bigg|_{\theta=0}$$

$$= \int_0^T \int_{\mathbb{R}^n} \int_{\mathbb{R}^n} \frac{\partial f}{\partial m}(x,m,\hat{v}(x))(\xi)\tilde{m}(\xi)m(x)dtd\xi dx$$

$$+ \int_0^T \int_{\mathbb{R}^n} f(x,m,\hat{v}(x))\tilde{m}(x)dtdx + \int_0^T \int_{\mathbb{R}^n} \frac{\partial f}{\partial v}(x,m,\hat{v}(x))v(x)m(x)dtdx$$

$$+ \int_{\mathbb{R}^n} h(x,m(T))\tilde{m}(x,T)dx + \int_{\mathbb{R}^n} \int_{\mathbb{R}^n} \frac{\partial h}{\partial m}(x,m(T))(\xi)\tilde{m}(\xi,T)m(x,T)d\xi dx.$$

$$\tag{4.5}$$

We introduce the function $u(x,t)$ solution of

$$-\frac{\partial u}{\partial t} + Au - g(x,m,\hat{v}(x)) \cdot Du - \int_{\mathbb{R}^n} Du(\xi) \cdot \frac{\partial g}{\partial m}(\xi,m,\hat{v}(\xi))(x)m(\xi)d\xi$$

$$= f(x,m,\hat{v}(x)) + \int_{\mathbb{R}^n} \frac{\partial f}{\partial m}(\xi,m,\hat{v}(\xi))(x)m(\xi)d\xi,$$

$$u(x,T) = h(x,m(T)) + \int_{\mathbb{R}^n} \frac{\partial h}{\partial m}(\xi,m(T))(x)m(\xi,T)d\xi \tag{4.6}$$

hence from (4.5)

$$\frac{dJ(\hat{v}(.)+\theta v(.), m_{\hat{v}(x)+\theta v(x)}(.))}{d\theta}\bigg|_{\theta=0}$$

$$= \int_0^T \int_{\mathbb{R}^n} \frac{\partial f}{\partial v}(x,m,\hat{v}(x))v(x)m(x)dtdx + \int_0^T \int_{\mathbb{R}^n} \left[-\frac{\partial u}{\partial t} + Au - g(x,m,\hat{v}(x)) \cdot Du\right.$$

$$\left. - \int_{\mathbb{R}^n} Du(\xi) \cdot \frac{\partial g}{\partial m}(\xi,m,\hat{v}(\xi))(x)m(\xi)d\xi\right]\tilde{m}(x)dtdx + \int_{\mathbb{R}^n} u(x,T)\tilde{m}(x,T))dx.$$

Using (4.4) we deduce

$$\frac{dJ(\hat{v}(.)+\theta v(.),m_{\hat{v}(x)+\theta v(x)}(.))}{d\theta}|_{\theta=0} = \int_0^T \int_{\mathbb{R}^n} \frac{\partial f}{\partial v}(x,m,\hat{v}(x))v(x)m(x)dtdx$$

$$- \int_0^T \int_{\mathbb{R}^n} u(x)\mathrm{div}(\frac{\partial g}{\partial v}(x,m,\hat{v}(x))v(x))m(x))dtdx$$

hence

$$\frac{dJ(\hat{v}(.)+\theta v(.),m_{\hat{v}(x)+\theta v(x)}(.))}{d\theta}|_{\theta=0} = \int_0^T \int_{\mathbb{R}^n} \frac{\partial f}{\partial v}(x,m,\hat{v}(x))v(x)m(x)dtdx$$

$$+ \int_0^T \int_{\mathbb{R}^n} Du(x)\cdot\frac{\partial g}{\partial v}(x,m,\hat{v}(x))v(x)m(x)dtdx.$$

Since $\hat{v}(.)$ is optimal, this expression must vanish for any $v(.)$. Hence necessarily

$$\frac{\partial f}{\partial v}(x,m,\hat{v}(x)) + \frac{\partial g}{\partial v}^*(x,m,\hat{v}(x))Du(x) = 0. \qquad (4.7)$$

It follows that (at least with convexity assumptions)

$$\hat{v}(x) = \hat{v}(x,m,Du(x)). \qquad (4.8)$$

We note that

$$f(x,m,\hat{v}(x)) + g(x,m,\hat{v}(x)),Du = H(x,m,Du) \qquad (4.9)$$

$$\int_{\mathbb{R}^n} \left[\frac{\partial f}{\partial m}(\xi,m,\hat{v}(\xi))(x) + Du(\xi)\cdot\frac{\partial g}{\partial m}(\xi,m,\hat{v}(\xi))(x)\right]m(\xi)d\xi$$

$$= \int_{\mathbb{R}^n} \frac{\partial H}{\partial m}(\xi,m,Du(\xi))(x)m(\xi)d\xi. \qquad (4.10)$$

We also note that

$$g(x,m,\hat{v}(x)) = g(x,m,\hat{v}(x,m,Du(x))) = G(x,m,Du). \qquad (4.11)$$

Going back to (4.2), written for $v(.) = \hat{v}(.)$, we can finally write the system of HJB-FP PDEs

$$-\frac{\partial u}{\partial t} + Au = H(x,m,Du) + \int_{\mathbb{R}^n} \frac{\partial H}{\partial m}(\xi,m,Du(\xi))(x)m(\xi)d\xi,$$

$$u(x,T) = h(x,m(T)) + \int_{\mathbb{R}^n} \frac{\partial h}{\partial m}(\xi,m(T))(x)m(\xi,T)d\xi,$$

$$\frac{\partial m}{\partial t} + A^*m + \mathrm{div}\ (G(x,m,Du)m) = 0,$$

$$m(x,0) = m_0(x). \tag{4.12}$$

We can compare the system (4.12) with (3.11). They differ through the partial derivative of H and h with respect to m. The optimal feedback is

$$\hat{v}(x) = \hat{v}(x,m,Du(x)). \tag{4.13}$$

Remark 3. Although they are similar, the systems (3.11) and (4.12) have been derived in a completely different manner. This is related to the time inconsistency issue. The derivation of (3.11) follows the standard pattern of dynamic programming, since m is external. This approach cannot be used to derive (4.12), which is obtained as a necessary condition of optimality.

The system (4.12) has not appeared in the literature, which relies on the stochastic maximum principle. This is natural, since we express merely necessary conditions of optimality.

4.2 Other Approaches

We first give another formula for the functional (4.3) and its Gateaux-derivative (4.5).

For a given feedback control $v(.)$, we introduce the linear equation

$$-\frac{\partial u_{v(.)}}{\partial t} + Au_{v(.)} - g(x,m_{v(.)}(t),v(x)) \cdot Du_{v(.)} = f(x,m_{v(.)}(t),v(x)),$$

$$u_{v(.)}(x,T) = h(x,m_{v(.)}(T)). \tag{4.14}$$

It is then easy to check, combining (4.2) and (4.14), that the functional (4.3) can be written as follows

$$J(v(.),m_{v(.)}(.)) = \int_{\mathbb{R}^n} u_{v(.)}(x,0)m_0(x)dx. \tag{4.15}$$

We can then give another expression for the Gateaux differential. We have considered the Gateaux differential [recall that $m_{\hat{v}(.)}(x) = m(x)$]

$$\frac{dm_{\hat{v}(.)+\theta v(.)}(x)}{d\theta}\Big|_{\theta=0} = \tilde{m}(x).$$

We can consider similarly

$$\frac{du_{\hat{v}(.)+\theta v(.)}(x)}{d\theta}\Big|_{\theta=0} = \tilde{u}(x)$$

and $\tilde{u}(x)$ is the solution of the linear equation

$$-\frac{\partial \tilde{u}}{\partial t} + A\tilde{u} - g(x,m(t),\hat{v}(x)).D\tilde{u}$$

$$-\left(\int \frac{\partial g}{\partial m}(x,m(t),\hat{v}(x))(\xi)\tilde{m}(\xi)d\xi + \frac{\partial g}{\partial v}(x,m(t),\hat{v}(x))v(x)\right) \cdot Du_{\hat{v}(.)}(x)$$

$$= \int \frac{\partial f}{\partial m}(x,m(t),\hat{v}(x))(\xi)\tilde{m}(\xi)d\xi + \frac{\partial f}{\partial v}(x,m(t),\hat{v}(x))v(x),$$

$$\tilde{u}(x,T) = \int \frac{\partial h}{\partial m}(x,m(T))(\xi)\tilde{m}(\xi,T)d\xi. \tag{4.16}$$

We can check the relation

$$\frac{dJ(\hat{v}(.) + \theta v(.), m_{\hat{v}(x)+\theta v(x)}(.))}{d\theta}\Big|_{\theta=0} = \int \tilde{u}(x,0)m_0(x)dx. \tag{4.17}$$

In fact, formulas (4.5) and (4.17) coincide. We leave it to the reader to check this.
    Computation of the Gateaux derivative allows us to write the optimality condition (4.7) as follows

$$\frac{\partial L}{\partial v}(x,m(t),\hat{v}(x),Du(x,t)) = 0,$$

where

$$L(x,m,v,q) = f(x,m,v) + q.g(x,m,v).$$

We have interpreted this condition as a necessary condition for $L(x,m(t),v,Du(x,t))$ to attain its minimum in v at $v = \hat{v}(x,t)$. This requires convexity.
    We can prove this minimum property directly, as in the proof of Pontryagin's maximum principle, by using spike changes for the optimal control.
    Suppose we change the optimum feedback $\hat{v}(x,s)$ into

$$\bar{v}(x,s) = \begin{vmatrix} v(x) & s \in (t,t+\varepsilon) \\ \hat{v}(x,s) & s \notin (t,t+\varepsilon) \end{vmatrix}$$

in which $v(x)$ is arbitrary (spike modification).

We then define $\bar{m} = m_{\bar{v}(.)}$. We have

$$\bar{m}(x,s) = m(x,s), \ \forall s \leq t,$$

$$\frac{\partial \bar{m}}{\partial t} + A^*\bar{m} + \text{div}\ (g(x,\bar{m},v(x))\bar{m}) = 0, \ t < s < t + \varepsilon,$$

$$\frac{\partial \bar{m}}{\partial t} + A^*\bar{m} + \text{div}\ (g(x,\bar{m},\hat{v}(x))\bar{m}) = 0, \ s > t + \varepsilon.$$

A tedious calculation allows us to write

$$J(\bar{v}(.),\bar{m}(.)) - J(\hat{v}(.),m(.))$$

$$= \int_t^{t+\varepsilon} \int_{\mathbb{R}^n} (L(x,m(s),v(x),Du) - L(x,m(s),\hat{v}(x),Du))m(x,s)dxds$$

$$+ \int_t^{t+\varepsilon} \int_{\mathbb{R}^n} (L(x,\bar{m}(s),v(x),Du)\bar{m}(x,s) - L(x,m(s),v(x),Du)m(x,s))dxds$$

$$- \int_t^{t+\varepsilon} \int_{\mathbb{R}^n} (\bar{m}(x,s) - m(x,s))(L(x,m(s),\hat{v}(x),Du)$$

$$+ \int \frac{\partial L}{\partial m}(\xi,m(s),\hat{v}(\xi),Du(\xi))(x)m(\xi,s)d\xi)dxds$$

$$+ \int_{t+\varepsilon}^T \int_{\mathbb{R}^n} (L(x,\bar{m}(s),\hat{v}(x),Du) - L(x,m(s),\hat{v}(x),Du))(\bar{m}(x,s)$$

$$- m(x,s))dxds + \int_{t+\varepsilon}^T \int_{\mathbb{R}^n} m(x,s)[L(x,\bar{m}(s),\hat{v}(x),Du) - L(x,m(s),\hat{v}(x),Du)$$

$$- \int \frac{\partial L}{\partial m}(x,m(s),\hat{v}(x),Du(x))(\xi)(\bar{m}(\xi,s) - m(\xi,s))d\xi]dxds$$

$$+ \int_{\mathbb{R}^n} (h(x,\bar{m}(T)) - h(x,m(T)))(\bar{m}(x,T) - m(x,T))dx$$

$$+ \int_{\mathbb{R}^n} m(x,T)[h(x,\bar{m}(T)) - h(x,m(T)) - \int \frac{\partial h}{\partial m}(x,m(T))(\xi)(\bar{m}(\xi,T) - m(\xi,T))d\xi]dx.$$

The left-hand side is positive. Dividing by ε and letting ε go to 0 yields

$$\int_{\mathbb{R}^n} (L(x,m(t),v(x),Du(x,t)) - L(x,m(t),\hat{v}(x,t),Du(x,t)))m(x,t)dx \geq 0, \ \text{a.e.}\ t$$

and since $v(x)$ is arbitrary, we get

$$L(x,m(t),v,Du(x,t)) \geq L(x,m(t),\hat{v}(x,t),Du(x,t)), \ \text{a.e.}\ x,t$$

which expresses the minimality property.

4.3   Stochastic Maximum Principle

We can derive from the system (4.12) a stochastic maximum principle. We proceed as in Sect. 3.2.

From the optimal feedback $\hat{v}(x)$ and the probability distribution $m(t)$ we construct stochastic processes $X(t) \in \mathbb{R}^n$, $V(t) \in \mathbb{R}^d$, $Y(t) \in \mathbb{R}^n$, and $Z(t) \in \mathcal{L}(\mathbb{R}^n; \mathbb{R}^n)$ which are adapted and defined as follows

$$X(t) = \hat{x}(t), \ m(t) = P_{X(t)}.$$

We next define

$$Y(t) = Du(X(t),t), \ V(t) = \hat{v}(X(t), P_{X(t)}, Y(t))$$

and finally

$$Z(t) = D^2 u \, \sigma(X(t), t).$$

We first have

$$dX = g(X(t), P_{X(t)}, V(t))dt + \sigma(X(t))dw(t)$$
$$X(0) = x_0. \tag{4.18}$$

We proceed as in Sect. 3.2 to obtain

$$-dY = \left(\frac{\partial f}{\partial x}(X(t), P_{X(t)}, V(t)) + \frac{\partial g}{\partial x}^*(X(t), P_{X(t)}, V(t))Y(t) + \text{tr} \, \frac{\partial \sigma(X(t))}{\partial x}^* Z(t) \right) dt$$

$$+ E \left[\frac{\partial^2 f}{\partial x \partial m}(X(t), P_{X(t)}, V(t)) + \frac{\partial^2 g}{\partial x \partial m}^*(X(t), P_{X(t)}, V(t))Y(t) \right] (X(t))dt$$

$$- Z(t)dw(t)$$

$$Y(T) = \frac{\partial h(X(T), P_{X(T)})}{\partial x} + E \left[\frac{\partial^2 h}{\partial x \partial m}(X(T), P_{X(T)}) \right] (X(T)).$$

The notation should be clearly understood to avoid confusion. When we write

$$E \left[\frac{\partial^2 f}{\partial x \partial m}(X(t), P_{X(t)}, V(t)) \right] (X(t))$$

we mean that we take the function $\frac{\partial f}{\partial m}(\xi, m, v)(x)$, where ξ and v are parameters and we take the gradient in x, denoted by $\frac{\partial^2 f}{\partial x \partial m}(\xi, m, v)(x)$. We then consider

$\xi = X(t)$, $v = V(t)$ and take the expected value $E\dfrac{\partial^2 f}{\partial x\partial m}(X(t),m,V(t))(x)$. We take $m = P_{X(t)}$ (note that it is a deterministic quantity) and thus get $E\dfrac{\partial^2 f}{\partial x\partial m}(X(t),P_{X(t)},V(t))(x)$. Finally, we take the argument $x = X(t)$. To emphasize the difficulty of confusion, consider $\dfrac{\partial f}{\partial x}(x,m.v)$. If we want to take the derivative with respect to m, then we should consider x,v as parameters, so change the notation to ξ and compute $\dfrac{\partial^2 f}{\partial m\partial x}(\xi,m,v)(x)$. Clearly

$$\frac{\partial^2 f}{\partial m\partial x}(\xi,m,v)(x) \neq \frac{\partial^2 f}{\partial x\partial m}(\xi,m,v)(x).$$

Recalling the definition of the Hamiltonian in this context, see (3.13), we can write the stochastic maximum principle as follows

$$X(t) \in R^n, V(t) \in R^d, Y(t) \in R^n, Z(t) \in \mathscr{L}(R^n;R^n)$$

$$dX = g(X(t),P_{X(t)},V(t))dt + \sigma(X(t))dw(t),$$

$$-dY = \left(\frac{\partial H}{\partial x}(X(t),P_{X(t)},V(t),Y(t)) + E\left[\frac{\partial^2 H}{\partial x\partial m}(X(t),P_{X(t)},V(t),Y(t))\right](X(t))\right.$$

$$\left.\times\mathrm{tr}\,\frac{\partial\sigma(X(t))}{\partial x}^{*}Z(t)\right)dt - Z(t)dw(t),$$

$$X(0) = x_0, \quad Y(T) = \frac{\partial h(X(T),P_{X(T)})}{\partial x} + E\left[\frac{\partial^2 h}{\partial x\partial m}(X(T),P_{X(T)})\right](X(T))$$

$$\text{(4.19)}$$

$$V(t)\text{ minimizes }\ H(X(t),P_{X(t)},v,Y(t))\text{ in }v. \tag{4.20}$$

Example 4. Consider again Example 2. We consider functions $f(x,\eta,v),g(x,\eta,v)$, $h(x,\eta)$, and

$$f(x,m,v) = f\left(x,\int_{\mathbb{R}^n}\varphi(\xi)m(\xi)d\xi,v\right),\ g(x,m,v) = f\left(x,\int_{\mathbb{R}^n}\chi(\xi)m(\xi)d\xi,v\right)$$

$$h(x,m) = h\left(x,\int_{\mathbb{R}^n}\psi(\xi)m(\xi)d\xi\right).$$

We get

$$\frac{\partial^2 f}{\partial x \partial m}(\xi,m,v)(x) = \frac{\partial \varphi^*}{\partial x}(x)\frac{\partial f}{\partial \eta}\left(\xi,\int_{\mathbb{R}^n}\varphi(\zeta)m(\zeta)d\zeta,v\right)$$

$$\frac{\partial^2 g^*}{\partial x \partial m}Du(\xi,m,v)(x) = \frac{\partial \chi^*}{\partial x}(x)\frac{\partial g^*}{\partial \eta}Du\left(\xi,\int_{\mathbb{R}^n}\chi(\zeta)m(\zeta)d\zeta,v\right)$$

$$\frac{\partial^2 h}{\partial x \partial m}(\xi,m)(x) = \frac{\partial \psi^*}{\partial x}(x)\frac{\partial h}{\partial \eta}\left(\xi,\int_{\mathbb{R}^n}\psi(\zeta)m(\zeta)d\zeta\right)$$

and the stochastic maximum principle writes

$$X(t) \in \mathbb{R}^n, V(t) \in \mathbb{R}^d, Y(t) \in \mathbb{R}^n, Z(t) \in \mathscr{L}(\mathbb{R}^n;\mathbb{R}^n)$$

$$dX = g(X(t),E\chi(X(t)),V(t))dt + \sigma(X(t))dw(t),$$

$$-dY = \left(\frac{\partial f}{\partial x}(X(t),E\varphi(X(t)),V(t)) + \frac{\partial g^*}{\partial x}(X(t),E\chi(X(t)),V(t))Y(t)\right.$$

$$+ \frac{\partial \varphi(X(t))^*}{\partial x}E\frac{\partial f}{\partial \eta}(X(t),E\varphi(X(t)),V(t))$$

$$+ \frac{\partial \chi(X(t))^*}{\partial x}E\left[\frac{\partial g^*}{\partial \eta}(X(t),E\chi(X(t)),V(t))Y(t)\right]$$

$$\left. \times \mathrm{tr}\frac{\partial \sigma(X(t))^*}{\partial x}Z(t)\right)dt - Z(t)dw(t),$$

$$X(0) = x_0, \quad Y(T) = \frac{\partial h(X(T),E\psi(X(T)))}{\partial x} + \frac{\partial \psi^*}{\partial x}(X(T))E\frac{\partial h}{\partial \eta}(X(T),E\psi(X(T)))$$

$$(4.21)$$

$$V(t) \text{ minimizes } f(X(t),E\varphi(X(t)),v) + Y(t).g(X(t),E\chi(X(t)),v) \text{ in } v. \quad (4.22)$$

We recover the stochastic maximum principle as established in [1, 12].

4.4 Time Inconsistency Approach

We discuss here the following particular mean field type problem

$$dx = g(x(t),v(x(t)))dt + \sigma(x(t))dw(t),$$

$$x(0) = x_0, \quad (4.23)$$

$$J(v(.),m(.)) = E\left[\int_0^T f(x(t),v(x(t))\,dt + h(x(T),m(T))\right]$$

$$+ \int_0^T F(Ex(t))dt + \Phi(Ex(T)). \tag{4.24}$$

We consider a feedback $v(x,t)$ and $m(t) = m_{v(.)}(t)$ is the probability density of $x_{v(.)}(t)$ the solution of (3.3). The functional becomes $J(v(.),m_{v(.)}(.))$. It is clearly a particular case of a mean field type control problem. We have indeed

$$f(x,m,v) = f(x,v) + F\left(\int \xi m(\xi)d\xi\right),$$

$$h(x,m) = h(x) + \Phi\left(\int \xi m(\xi)d\xi\right).$$

Therefore

$$H(x,m,q) = H(x,q) + F\left(\int \xi m(\xi)d\xi\right),$$

where

$$H(x,q) = \inf_v(f(x,v) + q.g(x,v)).$$

Considering $\hat{v}(x,q)$ which attains the infimum in the definition of $H(x,q)$ and setting

$$G(x,q) = g(x,\hat{v}(x,q))$$

the coupled system HJB-FP becomes, see (4.12),

$$-\frac{\partial u}{\partial t} + Au = H(x,Du) + F\left(\int \xi m(\xi)d\xi\right) + \sum_k \frac{\partial F}{\partial x_k}\left(\int \xi m(\xi)d\xi\right)x_k,$$

$$u(x,T) = h(x) + \Phi\left(\int \xi m(\xi)d\xi\right) + \sum_k \frac{\partial \Phi}{\partial x_k}\left(\int \xi m(\xi)d\xi\right)x_k,$$

$$\frac{\partial m}{\partial t} + A^*m + \mathrm{div}\,(G(x,Du)m) = 0,$$

$$m(x,0) = \delta(x-x_0). \tag{4.25}$$

We can reduce this problem slightly, using the following step: introduce the vector function $\Psi(x,t;s)$, $t < s$, which is the solution of

$$-\frac{\partial \Psi}{\partial t} + A\Psi - D\Psi.G(x,Du) = 0,\ t < s$$

$$\Psi(x,s;s) = x \tag{4.26}$$

then

$$\int \xi m(\xi,t)d\xi = \Psi(x_0,0;t)$$

so (4.25) becomes

$$-\frac{\partial u}{\partial t} + Au = H(x,Du) + F(\Psi(x_0,0;t)) + \sum_k \frac{\partial F}{\partial x_k}(\Psi(x_0,0;t))x_k$$

$$u(x,T) = h(x) + \Phi(\Psi(x_0,0;T)) + \sum_k \frac{\partial \Phi}{\partial x_k}(\Psi(x_0,0;T))x_k. \qquad (4.27)$$

We now have the system (4.26), (4.27). We can also look at $u(x,t)$ as the solution of a nonlocal HJB equation, depending on the initial state x_0. The optimal feedback

$$\hat{v}(x,t) = \hat{v}(x,Du(x,t))$$

depends also on x_0. Note that it does not depend on any intermediate state. This type of optimal control is called a *precommitment* optimal control.

In [11], the authors introduce a new concept, in order to define an optimization problem among feedbacks that do not depend on the initial condition. A feedback will be optimal only against spike changes, but not against global changes. They interpret this limited optimality as a game. Players are attached to small periods of time (eventually to each time, in the limit). Therefore, if one uses the concept of Nash equilibrium, decisions at different times correspond to decisions of different players, and thus are out of reach. This explains why only spike changes are allowed. Of course, in standard control problems, this is not a limitation, but it is one in the present situation.

In the spirit of dynamic programming, and the invariant embedding idea, we consider a family of control problems indexed by the initial conditions, and we control the system using feedbacks only. So if $v(x,s)$ is a feedback, we consider the state equation $x(s) = x_{xt}(s;v(.))$

$$dx = g(x(s),v(x(s),s))ds + \sigma(x(s))dw(t)$$

$$x(t) = x \qquad (4.28)$$

and the payoff

$$J_{x,t}(v(.)) = E\left[\int_t^T f(x(s),v(x(s),s))\,ds + h(x(T))\right]$$

$$+ \int_t^T F(Ex(s))ds + \Phi(Ex(T)). \qquad (4.29)$$

Consider a specific control $\hat{v}(x,s)$ that will be optimal in the sense described earlier. We define $\hat{x}(.)$ to be the corresponding state, the solution of (4.28), and set

$$V(x,t) = J_{x,t}(\hat{v}(.)). \qquad (4.30)$$

We make a spike modification and define

$$\bar{v}(x,s) = \begin{cases} v & t < s < t+\varepsilon \\ \hat{v}(x,s) & s > t+\varepsilon, \end{cases}$$

where v is arbitrary. The idea is to evaluate $J_{x,t}(\bar{v}(.))$ and to express that it is larger than $V(x,t)$. We define $\bar{x}(s)$ to the state corresponding to the feedback $\bar{v}(.)$. We have, by definition,

$$\bar{x}(s) = \begin{cases} x_{xt}(s;v), & t < s < t+\varepsilon \\ \hat{x}_{x_{xt}(t+\varepsilon;v),t+\varepsilon}(s), & t+\varepsilon < s < T, \end{cases}$$

where we have made explicit the initial conditions. In the sequel, to simplify the notation, we write

$$x(t+\varepsilon) = x_{xt}(t+\varepsilon;v).$$

We introduce the function

$$\Psi(x,t;s) = E\hat{x}_{xt}(s), \ t < s$$

which is the solution of

$$-\frac{\partial\Psi}{\partial t} + A\Psi - D\Psi \cdot g(x,\hat{v}(x,t)) = 0, \ t < s$$

$$\Psi(x,s;s) = x. \qquad (4.31)$$

We note the important property

$$E\bar{x}(s) = E\Psi(x(t+\varepsilon),t+\varepsilon;s), \ \forall s \geq t+\varepsilon.$$

Therefore.

$$J_{x,t}(\bar{v}(.)) = E\left[\int_t^{t+\varepsilon} f(x(s),v)ds + \int_{t+\varepsilon}^T f(\hat{x}_{x(t+\varepsilon),t+\varepsilon}(s),\hat{v}(\hat{x}_{x(t+\varepsilon),t+\varepsilon}(s),s))ds \right.$$

$$\left. + h(\hat{x}_{x(t+\varepsilon),t+\varepsilon}(T))\right] + \int_t^{t+\varepsilon} F(Ex(s))ds$$

$$+ \int_{t+\varepsilon}^T F(E\Psi(x(t+\varepsilon),t+\varepsilon;s))ds + \Phi(E\Psi(x(t+\varepsilon),t+\varepsilon;T)).$$

The next point is to compare $F(E\Psi(x(t+\varepsilon),t+\varepsilon;s))$ with $EF(\Psi(x(t+\varepsilon),t+\varepsilon;s))$. This is a simple application of Ito's formula

$$
EF(\Psi(x(t+\varepsilon),t+\varepsilon;s)
$$

$$
= F(\Psi(x,t;s)) + E\int_t^{t+\varepsilon}\sum_{ik}\frac{\partial F}{\partial x_k}(\Psi(x(\tau),\tau;s))\frac{\partial\Psi_k}{\partial x_i}(x(\tau),\tau;s)g_i(x(\tau),v)d\tau
$$

$$
+ E\int_t^{t+\varepsilon}\sum_{ij}a_{ij}(x(\tau))\left[\sum_{kl}\frac{\partial^2 F}{\partial x_k\partial x_l}(\Psi(x(\tau),\tau;s))\frac{\partial\Psi_k}{\partial x_i}\frac{\partial\Psi_l}{\partial x_j}(x(\tau),\tau;s)\right.
$$

$$
+ \sum_k\frac{\partial F}{\partial x_k}(\Psi(x(\tau),\tau;s))\frac{\partial^2\Psi_k}{\partial x_i\partial x_j}(x(\tau),\tau;s)\Bigg]d\tau
$$

$$
+ E\int_t^{t+\varepsilon}\sum_k\frac{\partial F}{\partial x_k}(\Psi(x(\tau),\tau;s))\frac{\partial\Psi_k}{\partial t}(x(\tau),\tau;s)d\tau.
$$

On the other hand,

$$
F(E\Psi(x(t+\varepsilon),t+\varepsilon;s))
$$

$$
= F(\Psi(x,t;s)) + \int_t^{t+\varepsilon}\sum_k\frac{\partial F}{\partial x_k}(E\Psi(x(\tau),\tau;s))E\left[\frac{\partial\Psi_k}{\partial t}(x(\tau),\tau;s)\right.
$$

$$
+ \sum_i\frac{\partial\Psi_k}{\partial x_i}(x(\tau),\tau;s)g_i(x(\tau),v) + \sum_{ij}a_{ij}(x(\tau))\frac{\partial^2\Psi_k}{\partial x_i\partial x_j}(x(\tau),\tau;s)\Bigg]d\tau.
$$

By comparison we have

$$
EF(\Psi(x(t+\varepsilon),t+\varepsilon;s)) - F(E\Psi(x(t+\varepsilon),t+\varepsilon;s))
$$

$$
= \varepsilon\sum_{ij}a_{ij}(x)\sum_{kl}\frac{\partial^2 F}{\partial x_k\partial x_l}(\Psi(x,t;s))\frac{\partial\Psi_k}{\partial x_i}\frac{\partial\Psi_l}{\partial x_j}(x,t;s)) + O(\varepsilon). \tag{4.32}
$$

We can similarly compute the difference $E\Phi(\Psi(x(t+\varepsilon),t+\varepsilon;T)) - \Phi(E\Psi(x(t+\varepsilon),t+\varepsilon;T))$. Collecting results, we obtain

$$
J_{x,t}(\bar{v}(.)) = EV(x(t+\varepsilon),t+\varepsilon) + \varepsilon\left[f(x,v) + F(x)\right.
$$

$$
- \sum_{ij}a_{ij}(x)\int_t^T\sum_{kl}\frac{\partial^2 F}{\partial x_k\partial x_l}(\Psi(x,t;s))\frac{\partial\Psi_k}{\partial x_i}\frac{\partial\Psi_l}{\partial x_j}(x,t;s))ds
$$

$$
- \sum_{ij}a_{ij}(x)\sum_{kl}\frac{\partial^2\Phi}{\partial x_k\partial x_l}(\Psi(x,t;T))\frac{\partial\Psi_k}{\partial x_i}\frac{\partial\Psi_l}{\partial x_j}(x,t;T))\Bigg] + O(\varepsilon).
$$

So we have the inequality

$$V(x,t) \leq EV(x(t+\varepsilon),t+\varepsilon) + \varepsilon \left[f(x,v) + F(x) \right.$$

$$- \sum_{ij} a_{ij}(x) \int_t^T \sum_{kl} \frac{\partial^2 F}{\partial x_k \partial x_l}(\Psi(x,t;s)) \frac{\partial \Psi_k}{\partial x_i} \frac{\partial \Psi_l}{\partial x_j}(x,t;s)ds$$

$$\left. - \sum_{ij} a_{ij}(x) \sum_{kl} \frac{\partial^2 \Phi}{\partial x_k \partial x_l}(\Psi(x,t;T)) \frac{\partial \Psi_k}{\partial x_i} \frac{\partial \Psi_l}{\partial x_j}(x,t;T) \right] + O(\varepsilon).$$

We expand $EV(x(t+\varepsilon),t+\varepsilon)$ by the same token, divide by ε and let $\varepsilon \to 0$. We obtain the inequality

$$0 \leq \frac{\partial V}{\partial t} + \sum_i \frac{\partial V}{\partial x_i} g_i(x,v) + \sum_{ij} a_{ij}(x) \frac{\partial^2 V}{\partial x_i \partial x_j} + f(x,v) + F(x)$$

$$- \sum_{ijkl} a_{ij}(x) \left[\int_t^T \frac{\partial^2 F}{\partial x_k \partial x_l}(\Psi(x,t;s)) \frac{\partial \Psi_k}{\partial x_i} \frac{\partial \Psi_l}{\partial x_j}(x,t;s)ds \right.$$

$$\left. + \frac{\partial^2 \Phi}{\partial x_k \partial x_l}(\Psi(x,t;T)) \frac{\partial \Psi_k}{\partial x_i} \frac{\partial \Psi_l}{\partial x_j}(x,t;T) \right].$$

Since we get an equality for $v = \hat{v}(x,t)$, we deduce easily

$$-\frac{\partial V}{\partial t} + AV = H(x,DV) + F(x) - \sum_{ijkl} a_{ij}(x) \left[\int_t^T \frac{\partial^2 F}{\partial x_k \partial x_l}(\Psi(x,t;s)) \frac{\partial \Psi_k}{\partial x_i} \frac{\partial \Psi_l}{\partial x_j}(x,t;s)ds \right.$$

$$\left. + \frac{\partial^2 \Phi}{\partial x_k \partial x_l}(\Psi(x,t;T)) \frac{\partial \Psi_k}{\partial x_i} \frac{\partial \Psi_l}{\partial x_j}(x,t;T) \right]$$

$$V(x,T) = h(x) + \Phi(x). \tag{4.33}$$

Moreover, the equation for Ψ can be written as

$$-\frac{\partial \Psi}{\partial t} + A\Psi - D\Psi \cdot G(x,DV) = 0, \ t < s$$

$$\Psi(x,s;s) = x \tag{4.34}$$

Remark 5. The optimal feedback obtained from the system (4.33), (4.34), is time consistent. It does not depend on the initial condition. The drawback is that we cannot extend this approach to more complex modelling, as we can in the mean field type treatment. On the other hand, this time consistent approach can be extended to situations in which the functionals depend on the initial conditions. For instance, we can consider instead of

$$Eh(x(T)) + \Phi(Ex(T))$$

the quantity

$$Eh(x,t;x(T)) + \Phi(x,t;Ex(T)).$$

Even when $\Phi = 0$, the problem is nonstandard.

Chapter 5
Approximation of Nash Games with a Large Number of Players

5.1 Preliminaries

We first assume that the functions $f(x,m,v)$, $g(x,m,v)$, and $h(x,m)$—as functions of m—can be extended to Dirac measures and the sum of Dirac measures that are probabilities. Since m is no more in L^p, the reference topology will be the weak $*$ topology of measures. That will be sufficient for our purpose, but we refer to [14] for metric space topology. At any rate, the vector space property is lost.

We consider N players. Each player is characterized by his or her state space. The state spaces are identical and are $\mathbb{R}^n$. We denote by $x^i(t)$ the state space of player i. The index of players will be an upper index. We write

$$x(t) = (x^1(t), \dots, x^N(t)).$$

The controls are feedbacks. An important limitation is that the feedbacks are limited to the individual states of players. So the feedback of player i is of the form $v^i(x^i)$. We will denote

$$v(x) = (v^1(x^1), \dots, v^N(x^N)).$$

We consider N independent Wiener processes $w^i(t)$, $i = 1, \dots, N$.

The trajectories are given by

$$dx^i = g\left(x^i(t), \frac{1}{N-1} \sum_{\substack{j=1 \neq i}}^{N} \delta_{x^j(t)}, v^i(x^i(t))\right) dt + \sigma(x^i(t)) dw^i(t)$$

$$x^i(0) = x^i_0. \tag{5.1}$$

We see that m has been replaced with $\frac{1}{N-1} \sum_{j=1 \neq i}^{N} \delta_{x^j}$ which is a probability on $\mathbb{R}^n$. The random variables x^i_0 are independent and identically distributed with density m_0. They are independent of the Wiener processes. The objective functional of player i is defined by

A. Bensoussan et al., *Mean Field Games and Mean Field Type Control Theory*,
SpringerBriefs in Mathematics, DOI 10.1007/978-1-4614-8508-7_5,
© Alain Bensoussan, Jens Frehse, Phillip Yam 2013

$$\mathcal{J}^i(v(.)) = E\left[\int_0^T f(x^i(t)), \frac{1}{N-1}\sum_{j=1\neq i}^N \delta_{x^j(t)}, v^i(x^i(t))dt\right.$$

$$\left. + h(x^i(T)), \frac{1}{N-1}\sum_{j=1\neq i}^N \delta_{x^j(T)}\right]. \tag{5.2}$$

5.2 System of PDEs

To any feedback control $v(.)$ we can associate a system of linear PDEs. Find functions $\Phi^i(x,t;v(.))$ that are the solution of

$$-\frac{\partial \Phi^i}{\partial t} + \sum_{h=1}^N A_{x^h}\Phi^i - \sum_{h=1}^N D_{x^h}\Phi^i \cdot g\left(x^h, \frac{1}{N-1}\sum_{j=1\neq h}^N \delta_{x^j}, v^h(x^h)\right)$$

$$= f\left(x^i, \frac{1}{N-1}\sum_{j=1\neq i}^N \delta_{x^j}, v^i(x^i)\right)$$

$$\Phi^i(x,T) = h\left(x^i, \frac{1}{N-1}\sum_{j=1\neq i}^N \delta_{x^j}\right), \tag{5.3}$$

where A_{x^h} refers to the differential operator A operating on the variable x^h. By simple application of Ito's formula, one can check easily that

$$\mathcal{J}^i(v(.)) = \int \cdots \int \Phi^i(x,0)\prod_{j=1}^N m_0(x^j)dx^j. \tag{5.4}$$

Remark 6. To find a Nash equilibrium, one can write a system of nonlinear PDE, as in [5]. However the feedback one can obtain in this way is global, and uses the state values of all players. So a Nash equilibrium related to individual state values will not exist in general. The idea of mean field games is that we can obtain an approximate Nash equilibrium based on individual states, which is good because N is large. Nevertheless, we will consider in the next section a particular case in which a Nash equilibrium among individual feedbacks exists.

5.3 Independent Trajectories

In this section, we assume that $g(x,m,v) = g(x,v)$ is independent of m. Then the individual trajectories become

$$dx^i = g(x^i(t), v^i(x^i(t))dt + \sigma(x^i(t))dw^i(t)$$

$$x^i(0) = x_0^i. \tag{5.5}$$

It is clear that the processes are now independent. The probability distribution of $x^i(t)$ is the function $m_{v^i(.)}(x^i,t)$ solution of the FP equation

$$\frac{\partial m_{v^i(.)}}{\partial t} + A_{x^i} m_{v^i(.)} + \operatorname{div}_{x^i}(g(x^i,v^i(x^i))m_{v^i(.)}(x^i)) = 0$$

$$m_{v^i(.)}(x^i,0) = m_0(x^i). \qquad (5.6)$$

We then define the function

$$\Psi^i(x^i,t) = \int \cdots \int \prod_{j \neq i} \Phi^i(x,t) m_{v^j(.)}(x^j,t) dx^j. \qquad (5.7)$$

By testing (5.3) with $\prod_{j \neq i} m_{v^j(.)}(x^j)$ and integrating, and using (5.6), we obtain

$$-\frac{\partial \Psi^i}{\partial t} + A_{x^i}\Psi^i - D_{x^i}\Psi^i \cdot g(x^i,v^i(x^i))$$

$$= \int \cdots \int f\left(x^i, \frac{1}{N-1}\sum_{h=1 \neq i}^{N} \delta_{x^h}, v^i(x^i)\right) \prod_{j \neq i} m_{v^j(.)}(x^j) dx^j$$

$$\Psi^i(x^i,T) = \int \cdots \int h\left(x^i, \frac{1}{N-1}\sum_{h=1 \neq i}^{N} \delta_{x^h}\right) \prod_{j \neq i} m_{v^j(.)}(x^j,T) dx^j. \qquad (5.8)$$

We also have

$$\int_{\mathbb{R}^n} \Psi^i(x^i,t) m_{v^i(.)}(x^i,t) dx^i = \int \cdots \int \Phi^i(x,t) \prod_{j} m_{v^j(.)}(x^j,t) dx^j. \qquad (5.9)$$

In particular

$$\mathcal{J}^i(v(.)) = \int_{\mathbb{R}^n} \Psi^i(x^i,0) m_0(x^i) dx^i. \qquad (5.10)$$

If we have a Nash equilibrium $\hat{v}(.)$, which we write $(\hat{v}^i(.),\overline{\hat{v}^i}(.))$ in which $\overline{\hat{v}^i}(.)$ represents all components different from i, then, noting $\Psi^i(x^i,t;v^i(.),\overline{\hat{v}^i}(.))$, the solution of (5.8) when we take $v^j(x^j) = \hat{v}^j(x^j), \forall j \neq i$, then we can assert that

$$\Psi^i(x^i,t;\hat{v}^i(.),\overline{\hat{v}^i}(.)) \leq \Psi^i(x^i,t;v^i(.),\overline{\hat{v}^i}(.)), \forall v^i(.).$$

From standard dynamic programming theory, we obtain that the functions

$$u^i(x^i,t) = \Psi^i(x,t;\hat{v}^i(.),\overline{\hat{v}^i}(.)) \qquad (5.11)$$

satisfy

$$-\frac{\partial u^i}{\partial t} + A_{x^i} u^i$$

$$= \inf_v \left[\int \cdots \int f\left(x^i, \frac{1}{N-1} \sum_{h=1\neq i}^{N} \delta_{x^h}, v \right) \prod_{j\neq i} m_{\hat{v}^j(.)}(x^j)dx^j + Du^i(x^i).g(x^i,v) \right].$$
(5.12)

We also have the terminal condition

$$u^i(x^i,T) = \int \cdots \int h\left(x^i, \frac{1}{N-1} \sum_{h=1\neq i}^{N} \delta_{x^h} \right) \prod_{j\neq i} m_{\hat{v}^j(.)}(x^j,T)dx^j.$$
(5.13)

We want to show that a Nash equilibrium exists, made of identical feedbacks for all players. We proceed as follows: For $x \in \mathbb{R}^n, m \in L^1(\mathbb{R}^n), q \in \mathbb{R}^n$ define the Hamiltonian

$$H_N(x,m,q) = \inf_v \left[\int \cdots \int f\left(x, \frac{1}{N-1} \sum_{h=1}^{N-1} \delta_{x^h}, v \right) \prod_{j=1}^{N-1} m(x^j)dx^j + q.g(x,v) \right] \quad (5.14)$$

and let $\hat{v}_N(x,m,q)$ be the minimizer in the Hamiltonian. Set

$$G_N(x,m,q) = g(x,\hat{v}_N(x,m,q)).$$
(5.15)

Define finally

$$h_N(x,m) = \int \cdots \int h\left(x, \frac{1}{N-1} \sum_{h=1}^{N-1} \delta_{x^h} \right) \prod_{j=1}^{N-1} m(x^j)dx^j.$$
(5.16)

We consider the pair of functions (if it exists) $u_N(x,t)$, $m_N(x,t)$ the solution of the system

$$-\frac{\partial u_N}{\partial t} + Au_N = H_N(x,m_N,Du_N)$$

$$\frac{\partial m_N}{\partial t} + A^* m_N + \mathrm{div}(G_N(x,m_N,Du_N)m_N) = 0$$

$$u_N(x,T) = h_N(x,m(T)), \ m_N(x,0) = m_0(x).$$
(5.17)

Then, from symmetry considerations, it is easy to check that the functions $u^i(x^i,t)$ coincide with $u_N(x^i,t)$. We can next make the connection with the differential game for N players. Write $\mathcal{J}^{N,i}(v(.))$ the objective functional of player i defined by (5.2), to emphasize that there are N players. Define common feedbacks

$$\hat{v}_N(x) = \hat{v}_N(x,m_N(x),u_N(x))$$
(5.18)

and denote by $\mathcal{J}^{N,i}(\hat{v}_N(.))$ the value of the objective functional of Player i, when all players use the same feedback. We can assert that

$$\mathcal{J}^{N,i}(\hat{v}_N(.)) = \int u_N(x,0)m_0(x)dx \qquad (5.19)$$

and when all players use the same local feedback $\hat{v}_N(.)$ one gets a Nash equilibrium for the functionals $\mathcal{J}^{N,i}(v(.))$ among local feedbacks (i.e., feedbacks on individual states).

In this case, mean field theory is concerned with what happens to the system (5.17) as $N \to +\infty$. To simplify the analysis, consider the situation

$$f(x,m,v) = f_0(x,m) + f(x,v) \qquad (5.20)$$

then

$$H_N(x,m,q) = \int \cdots \int f_0\left(x, \frac{1}{N-1}\sum_{h=1}^{N-1}\delta_{x^h}\right)\prod_{j=1}^{N-1}m(x^j)dx^j + H(x,q) \qquad (5.21)$$

in which

$$H(x,q) = \inf_v[f(x,v) + q.g(x,v)] \qquad (5.22)$$

and $\hat{v}_N(x,m,q) = \hat{v}(x,q)$, which is the minimizer in (5.22). Also

$$G_N(x,m,q) = G(x,q) = g(x,\hat{v}(x,q)).$$

If we set

$$f_{0N}(x,m) = \int \cdots \int f_0\left(x, \frac{1}{N-1}\sum_{h=1}^{N-1}\delta_{x^h}\right)\prod_{j=1}^{N-1}m(x^j)dx^j \qquad (5.23)$$

then the system (5.17) amounts to

$$-\frac{\partial u_N}{\partial t} + Au_N = H(x,Du_N) + f_{0N}(x,m)$$

$$\frac{\partial m_N}{\partial t} + A^*m_N + \mathrm{div}(G(x,Du_N)m_N) = 0$$

$$u_N(x,T) = h_N(x,m(T)), \quad m_N(x,0) = m_0(x). \qquad (5.24)$$

It is clear that the pair $u_N(x,t), m_N(x,t)$ will converge pointwise and in Sobolev spaces towards the $u(x,t), m(x,t)$ solution of

$$-\frac{\partial u}{\partial t} + Au = H(x, Du) + f_0(x, m)$$

$$\frac{\partial m}{\partial t} + A^* m + \operatorname{div}(G(x, Du)m) = 0$$

$$u(x, T) = h(x, m(T)), \ m(x, 0) = m_0(x) \tag{5.25}$$

provided $f_{0N}(x, m(t)) \to f_0(x, m(t))$ and $h_N(x, m(T)) \to h(x, m(T))$ as $N \to +\infty$, for any fixed x, t. This is a consequence of the law of large numbers. Indeed, consider a sequence of independent random variables, identically distributed with probability distribution $m(x)$ (this is done for any t, so we do not mention the time dependence). We denote this sequence by X^j. Then

$$f_{0N}(x, m) = E f_0 \left(x, \frac{1}{N-1} \sum_{h=1}^{N-1} \delta_{X^h} \right). \tag{5.26}$$

We claim that the random measure on $\mathbb{R}^n$, $\frac{1}{N-1} \sum_{h=1}^{N-1} \delta_{X^h}$ converges a.s. towards m, for the weak $*$ topology of measures on $\mathbb{R}^n$. Indeed, consider a continuous bounded function on $\mathbb{R}^n$, denoted by φ; then

$$\left\langle \varphi, \frac{1}{N-1} \sum_{h=1}^{N-1} \delta_{X^h} \right\rangle = \frac{1}{N-1} \sum_{h=1}^{N-1} \varphi(X^h).$$

But the real random variables $\varphi(X^h)$ are independent and identically distributed. The mean is $\int \varphi(x) m(x) dx$. According to the law of large numbers

$$\frac{1}{N-1} \sum_{h=1}^{N-1} \varphi(X^h) \to \int \varphi(x) m(x) dx, \ \text{a.s. as } N \to +\infty$$

hence the convergence of $\frac{1}{N-1} \sum_{h=1}^{N-1} \delta_{X^h}$ towards m, a.s. as $N \to +\infty$. If we assume that $f_0(x, m)$ is continuous in m for the topology of weak $*$ convergence, we get

$$f_0 \left(x, \frac{1}{N-1} \sum_{h=1}^{N-1} \delta_{X^h} \right) \to f_0(x, m) \ \text{a.s. as } N \to +\infty$$

and $f_{0N}(x, m) \to f_0(x, m)$, provided Lebesgue's theorem can be applied.

5.4 General Case

In the general case, namely if g depends on m as in (5.1), (5.2), the problem of Nash equilibrium among local feedbacks (i.e., those depending on individual states) has no solution. In that case, the mean field theory can provide a good feedback control, when N is large. However, PDE techniques cannot be used, since the Bellman system necessitates allowing global feedbacks (i.e., those based on the states of all players). The approximation property can be shown only with probability techniques. In this context, there is no reason to consider specific feedbacks, largely motivated by PDE techniques. The simplest is to consider open loop local controls. These are random processes adapted to the individual Wiener processes (the uncertainty that affects the evolution of the state of each individual player), but not linked to the state. So we reformulate the game as follows

$$dx^i = g(x^i(t), \frac{1}{N-1} \sum_{j=1 \neq i}^{N} \delta_{x^j(t)}, v^i(t)) dt + \sigma(x^i(t)) dw^i(t)$$

$$x^i(0) = x_0^i \tag{5.27}$$

$$\mathcal{J}^{N,i}(v(.)) = E \left[\int_0^T f \left(x^i(t), \frac{1}{N-1} \sum_{j=1 \neq i}^{N} \delta_{x^j(t)}, v^i(t) \right) dt \right.$$

$$\left. + h \left(x^i(T), \frac{1}{N-1} \sum_{j=1 \neq i}^{N} \delta_{x^j(T)} \right) \right]. \tag{5.28}$$

We recall that the Wiener processes $w^i(t)$ are standard one and independent, the x_0^i are independent identically distributed random variables, independent from the Wiener processes and with probability distribution $m_0(x)$. The processes $v^i(t)$ are random processes adapted to the filtration generated by $w^i(t)$, for each i. So when we write $\mathcal{J}^{N,i}(v(.))$, $v(.)$ should be understood as $(v^1(t), \ldots, v^N(t))$. As already mentioned there is no exact Nash equilibrium, among open loop controls. We now construct our approximation.

We go back to (2.3), (2.5). We consider the solution (3.1), (2.8). We consider the optimal feedback $\hat{v}(x) = \hat{v}(x, m, Du)$ where the pair u, m is the solution of the system (3.11) of coupled HJB-FP equations. We then construct the optimal trajectory

$$d\hat{x} = g(\hat{x}(t), m(t), \hat{v}(\hat{x}(t))) dt + \sigma(\hat{x}(t)) dw(t)$$

$$\hat{x}(0) = x_0 \tag{5.29}$$

and define the stochastic process $\hat{v}(t) = \hat{v}(\hat{x}(t))$, which is adapted to the pair $x_0, w(.)$. If we consider any process $v(t)$ adapted to the filtration generated by $x_0, w(.)$ and the trajectory

$$dx = g(x, m(t), v(t)) dt + \sigma(x) dw(t)$$

$$x(0) = x_0 \tag{5.30}$$

with the same $m(t)$ as in (5.29). The fact that $m(t)$ is the probability distribution of $\hat{x}(t)$ does not play a role in this aspect. Consider then the payoff functional

$$J(v(.)) = E\left[\int_0^T f(x(t),m(t),v(t))\,dt + h(x(T),m(T))\right] \qquad (5.31)$$

in which $v(.)$ refers to $v(t)$. From standard control theory, it follows that

$$\inf_{v(.)} J(v(.)) = J(\hat{v}(.)) = \int_{\mathbb{R}^n} u(x,0)m_0(x)dx. \qquad (5.32)$$

We now construct N replicas of $x_0, w(.)$ called $x_0^i, w^i(.)$, which are independent. We can then define N replicas of $\hat{v}(t)$, called $\hat{v}^i(t)$. Since $\hat{v}^i(t)$ is adapted to $x_0^i, w^i(.)$, they are independent processes. We can deduce N replicas of $\hat{x}(t)$, called $\hat{x}^i(t)$. They satisfy

$$d\hat{x}^i = g(\hat{x}^i(t),m(t),\hat{v}^i(t))dt + \sigma(\hat{x}^i)dw^i(t)$$

$$\hat{x}^i(0) = x_0^i. \qquad (5.33)$$

From their construction, it can be seen they are independent processes. Moreover,

$$J(\hat{y}^i(.)) = J(\hat{v}(.)). \qquad (5.34)$$

We will use $\hat{v}^i(t)$ in the context of the differential game (5.27), (5.28). We will show that they constitute an approximate local Nash equilibrium. We first compute $\mathcal{J}^{N,i}(\hat{v}(.))$. Note that, in the game (5.27), (5.28), the trajectories corresponding to the controls $\hat{v}^i(t)$ are not $\hat{x}^i(t)$. We call them $\hat{y}^i(t)$ and they are defined by

$$d\hat{y}^i = g\left(\hat{y}^i(t), \frac{1}{N-1}\sum_{j=1\neq i}^N \delta_{\hat{y}^j(t)}, v^i(t)\right)dt + \sigma(\hat{y}^i(t))dw^i(t)$$

$$\hat{y}^i(0) = x_0^i. \qquad (5.35)$$

Note that $\hat{y}^i(t)$ depends on N, which is not the case for $\hat{x}^i(t)$.

The first task is to show that $\hat{x}^i(t)$ constitutes a good approximation of $\hat{y}^i(t)$. We evaluate $\hat{y}^i(t) - \hat{x}^i(t)$ as follows

$$d(\hat{y}^i(t)-\hat{x}^i(t)) = \left(g\left(\hat{y}^i(t),\frac{1}{N-1}\sum_{j=1\neq i}^N \delta_{\hat{y}^j(t)}, v^i(t)\right) - g\left(\hat{x}^i(t),\frac{1}{N-1}\sum_{j=1\neq i}^N \delta_{\hat{x}^j(t)}, v^i(t)\right)\right)dt$$

$$+ \left(g\left(\hat{x}^i(t),\frac{1}{N-1}\sum_{j=1\neq i}^N \delta_{\hat{x}^j(t)}, v^i(t)\right) - g(\hat{x}^i(t),m(t),\hat{v}^i(t))\right)dt$$

$$+ (\sigma(\hat{y}^i(t)) - \sigma(\hat{x}^i(t)))dw^i(t)$$

$$\hat{y}^i(0) - \hat{x}^i(0) = 0.$$

It is clear from this expression that one needs Lipschitz assumptions for g and σ in x. An important issue is to evaluate

$$g\left(\hat{x}^i(t), \frac{1}{N-1}\sum_{j=1\neq i}^{N}\delta_{\hat{y}^j(t)}, v^i(t)\right) - g\left(\hat{x}^i(t), \frac{1}{N-1}\sum_{j=1\neq i}^{N}\delta_{\hat{x}^j(t)}, v^i(t)\right)$$

so we have to evaluate $g(x,\mu,v) - g(x,v,v)$ when $\mu = \frac{1}{N-1}\sum_{j=1}^{N-1}\delta_{\xi^j}$ and $v = \frac{1}{N-1}\sum_{j=1}^{N-1}\delta_{\eta^j}$, where ξ^j and η^j are points in $\mathbb{R}^n$. Assumptions must be made to estimate this difference with $\frac{1}{N-1}\sum_{j=1}^{N-1}|\xi^j - \eta^j|$. A standard case will be a function $g(x,\mu)$ of the form (omitting to indicate v)

$$g(x,\mu) = \varphi\left(\int K(x,y)\mu(dy)\right),$$

where φ is Lipschitz and $K(x,y)$ is Lipschitz in the second argument. There remains the driving term (not depending on $\hat{y}^i(t)$)

$$g(\hat{x}^i(t), \frac{1}{N-1}\sum_{j=1\neq i}^{N}\delta_{\hat{x}^j(t)}, v^i(t)) - g(\hat{x}^i(t), m(t), \hat{v}^i(t)).$$

But, since the random variables $\hat{x}^j(t)$ are independent and identically distributed with $m(t)$ for probability distribution, the random measure on R^n defined by $\frac{1}{N-1}\sum_{j=1\neq i}^{N}\delta_{\hat{x}^j(t)}$ converges a.s. to $m(t)$ for the weak * topology of measures. If $g(x,m,v)$ is continuous in m for the weak * topology of measures, we can assert that

$$g(\hat{x}^i(t), \frac{1}{N-1}\sum_{j=1\neq i}^{N}\delta_{\hat{x}^j(t)}, v^i(t)) - g(\hat{x}^i(t), m(t), \hat{v}^i(t)) \to 0, \text{ a.s.}$$

This sequence of steps, with appropriate assumptions shows that $\hat{y}^i(t) - \hat{x}^i(t) \to 0$ for instance in $L(0,T;L^2(\Omega,\mathcal{A},P))$ and $\hat{y}^i(T) - \hat{x}^i(T) \to 0$ in $L^2(\Omega,\mathcal{A},P)$. With this in hand, we can consider

$$\mathcal{J}^{N,i}(\hat{v}(.)) = E\left[\int_0^T f(\hat{y}^i(t), \frac{1}{N-1}\sum_{j=1\neq i}^{N}\delta_{\hat{y}^j(t)}, \hat{v}^i(t))dt\right.$$

$$\left. + h(\hat{y}^i(T), \frac{1}{N-1}\sum_{j=1\neq i}^{N}\delta_{\hat{y}^j(T)})\right].$$

We can write

$$\mathcal{J}^{N,i}(\hat{v}(.)) = J(\hat{v}^i(.)) + E\left[\int_0^T \left(f\left(\hat{y}^i(t), \frac{1}{N-1}\sum_{j=1\neq i}^N \delta_{\hat{y}^j(t)}, \hat{v}^i(t)\right) \right. \right.$$

$$\left. \left. -f\left(\hat{x}^i(t), \frac{1}{N-1}\sum_{j=1\neq i}^N \delta_{\hat{x}^j(t)}, \hat{v}^i(t)\right) \right) dt\right]$$

$$+ E\left[\int_0^T \left(f\left(\hat{x}^i(t), \frac{1}{N-1}\sum_{j=1\neq i}^N \delta_{\hat{x}^j(t)}, \hat{v}^i(t)\right) - f\left(\hat{x}^i(t), m(t), \hat{v}^i(t)\right) \right) dt\right]$$

$$+ E\left[h\left(\hat{y}^i(T), \frac{1}{N-1}\sum_{j=1\neq i}^N \delta_{\hat{y}^j(T)}\right) - h\left(\hat{x}^i(T), \frac{1}{N-1}\sum_{j=1\neq i}^N \delta_{\hat{x}^j(T)}\right)\right]$$

$$+ E\left[h\left(\hat{x}^i(T), \frac{1}{N-1}\sum_{j=1\neq i}^N \delta_{\hat{x}^j(T)}\right) - h(\hat{x}^i(T), m(T))\right].$$

Since $\hat{x}^i(t)$ approximates $\hat{y}^i(t)$, by appropriate smoothness assumptions on f and h, analogous to g, we can state that

$$\mathcal{J}^{N,i}(\hat{v}(.)) = J(\hat{v}^i(.)) + O\left(\frac{1}{\sqrt{N}}\right). \tag{5.36}$$

Let us now focus on player 1, but we can naturally take any player. Player 1 will use a different local control denoted by $v^1(t)$, and all other players use $\hat{v}^i(t), i \geq 2$. We associate to these controls the trajectories

$$dx^1 = g(x^1, m(t), v^1(t))dt + \sigma(x^1)dw^1(t)$$

$$x^1(0) = x_0^1 \tag{5.37}$$

and $\hat{x}^i(t), i \geq 2$. Call $\tilde{v}(.) = (v^1(.), \hat{v}^2(.), \dots \hat{v}^N(.))$ and use these controls in the differential game (5.27), (5.28). The corresponding states are denoted by $y^1(.), \dots, y^N(.)$ and are defined by the equations

$$dy^1 = g\left(y^1(t), \frac{1}{N-1}\sum_{j=2}^N \delta_{y^j(t)}, v^1(t)\right) dt + \sigma(y^1(t))dw^1(t)$$

$$y^1(0) = x_0^1 \tag{5.38}$$

$$dy^i = g\left(y^i(t), \frac{1}{N-1}\sum_{j=1\neq i}^N \delta_{y^j(t)}, \hat{v}^i(t)\right) dt + \sigma(y^i(t))dw^i(t)$$

$$y^i(0) = x_0^i. \tag{5.39}$$

We want to show again that $x^1(t)$ approximates $y^1(t)$ and $\hat{x}^i(t)$ approximates $y^i(t), \forall i \geq 2$. We begin with $i \geq 2$ and write

$$d(y^i(t) - \hat{x}^i(t)) = \left(g\left(y^i(t), \frac{1}{N-1} \sum_{j=1\neq i}^{N} \delta_{y^j(t)}, \hat{v}^i(t) \right) - g\left(y^i(t), \frac{1}{N-2} \sum_{j=2\neq i}^{N} \delta_{y^j(t)}, \hat{v}^i(t) \right) \right) dt$$

$$+ \left(g\left(y^i(t), \frac{1}{N-2} \sum_{j=2\neq i}^{N} \delta_{y^j(t)}, \hat{v}^i(t) \right) - g\left(\hat{x}^i(t), \frac{1}{N-2} \sum_{j=2\neq i}^{N} \delta_{\hat{x}^j(t)}, \hat{v}^i(t) \right) \right) dt$$

$$+ \left(g\left(\hat{x}^i(t), \frac{1}{N-2} \sum_{j=2\neq i}^{N} \delta_{\hat{x}^j(t)}, \hat{v}^i(t) \right) - g(\hat{x}^i(t), m(t), \hat{v}^i(t)) \right) dt$$

$$+ (\sigma(y^i(t)) - \sigma(\hat{x}^i(t)))dw^i(t)$$

$$y^i(0) - \hat{x}^i(0) = 0.$$

For the first term, assumptions must be made so that it is estimated by $\frac{\sum_{j=2}^{N}|y^1(t)-y^j(t)|}{(N-1)(N-2)}$, which should tend to 0 in $L^2(0,T;L^2(\Omega,\mathcal{A},P))$. It is sufficient to show that all processes are bounded in $L^2(0,T;L^2(\Omega,\mathcal{A},P))$. The other terms are dealt with as explained for $\hat{y}^i(t) - \hat{x}^i(t)$ above. This argument shows that $\hat{x}^i(t)$ approximates $y^i(t), \forall i \geq 2$. We next consider

$$d(y^1(t) - x^1(t)) = \left(g\left(y^1(t), \frac{1}{N-1} \sum_{j=2}^{N} \delta_{y^j(t)}, v^1(t) \right) - g\left(y^1(t), \frac{1}{N-1} \sum_{j=2}^{N} \delta_{\hat{x}^j(t)}, v^1(t) \right) \right) dt$$

$$+ \left(g\left(y^1(t), \frac{1}{N-1} \sum_{j=2}^{N} \delta_{\hat{x}^j(t)}, v^1(t) \right) - g\left(x^1(t), \frac{1}{N-1} \sum_{j=2}^{N} \delta_{\hat{x}^j(t)}, v^1(t) \right) \right) dt$$

$$+ \left(g\left(x^1(t), \frac{1}{N-1} \sum_{j=2}^{N} \delta_{\hat{x}^j(t)}, v^1(t) \right) - g(x^1, m(t), v^1(t)) \right) dt$$

$$+ \sigma(y^1(t)) - \sigma(x^1(t)))dw^1(t)$$

$$y^1(0) - x^1(0) = 0$$

and similar considerations as above show that $x^1(t)$ approximates $y^1(t)$. We then write

$$\mathcal{J}^{N,1}(\tilde{v}(.)) = E\left[\int_0^T f(y^1(t), \frac{1}{N-1} \sum_{j=2}^{N} \delta_{y^j(t)}, v^1(t))dt \right.$$

$$\left. + h\left(y^1(T), \frac{1}{N-1} \sum_{j=2}^{N} \delta_{y^j(T)} \right) \right].$$

Thus,

$$\mathcal{J}^{N,1}(\tilde{v}(.)) = J(v^1(.)) + E\left[\int_0^T \left(\left(y^1(t), \frac{1}{N-1}\sum_{j=2}^N \delta_{y^j(t)}, v^1(t)\right)\right.\right.$$

$$\left.\left. -f\left(x^1(t), \frac{1}{N-1}\sum_{j=2}^N \delta_{\hat{x}^j(t)}, v^1(t)\right)\right)dt\right]$$

$$+ E\left[\int_0^T \left(f\left(x^1(t), \frac{1}{N-1}\sum_{j=2}^N \delta_{\hat{x}^j(t)}, v^1(t)\right) - f\left(x^1(t), m(t), v^1(t)\right)\right)dt\right]$$

$$+ E\left[h\left(y^1(T), \frac{1}{N-1}\sum_{j=2}^N \delta_{y^j(T)}\right) - h\left(x^1(T), \frac{1}{N-1}\sum_{j=2}^N \delta_{\hat{x}^j(T)}\right)\right]$$

$$+ E\left[h\left(x^1(T), \frac{1}{N-1}\sum_{j=2}^N \delta_{\hat{x}^j(T)}\right) - h(x^1(T), m(T))\right].$$

Since $x^1(t)$ approximates $y^1(t)$ and $\hat{x}^i(t)$ approximates $y^i(t), \forall i \geq 2$, and since

$$E\left[h\left(x^1(T), \frac{1}{N-1}\sum_{j=2}^N \delta_{\hat{x}^j(T)}\right) - h(x^1(T), m(T))\right] \to 0$$

we can assert that

$$\mathcal{J}^{N,1}(\tilde{v}(.)) = J(v^1(.)) + O\left(\frac{1}{\sqrt{N}}\right).$$

By the standard control theory, $J(v^1(.)) \geq J(\hat{v}^1(.))$, therefore,

$$\mathcal{J}^{N,1}(\tilde{v}(.)) \geq \mathcal{J}^{N,1}(\hat{v}(.)) - O\left(\frac{1}{\sqrt{N}}\right)$$

which proves that $\hat{v}(.)$ is an approximate Nash equilibrium among local open loop controls.

5.5 Nash Equilibrium Among Local Feedbacks

In the preceding section, we considered local open loop controls. However the approximate Nash equilibrium that has been obtained is obtained through a feedback. So it is natural to ask whether we can obtain an approximate Nash equilibrium among local feedbacks (feedbacks on individual states).

Moreover, the feedback was given by $\hat{v}(x) = \hat{v}(x, m, Du)$, so we can consider that it is a Lipschitz function. So it is natural to consider the class of local feedbacks that are Lipschitz continuous. In that case, it is true that $\hat{v}(.) = (\hat{v}^1(.), \ldots, \hat{v}^N(.))$ constitutes an approximate Nash equilibrium.

Indeed, we consider the trajectories

$$d\hat{x}^i = g(\hat{x}^i(t), m(t), \hat{v}^i(\hat{x}^i(t)))dt + \sigma(\hat{x}^i)dw^i(t)$$

$$\hat{x}^i(0) = x_0^i. \tag{5.40}$$

Now if we use the feedbacks $\hat{v}(.)$ in the differential game (5.1), (5.2), we get the trajectories

$$d\hat{y}^i = g\left(\hat{y}^i(t), \frac{1}{N-1} \sum_{\substack{j=1 \\ j \neq i}}^{N} \delta_{\hat{y}^j(t)}, \hat{v}^i(\hat{y}^i(t))\right) dt + \sigma(\hat{y}^i(t))dw^i(t)$$

$$\hat{y}^i(0) = x_0^i. \tag{5.41}$$

We want to show that $\hat{x}^i(t)$ approximates $\hat{y}^i(t)$. The difference with the open loop case is that $\hat{v}^i(\hat{x}^i(t))$ and $\hat{v}^i(\hat{y}^i(t))$ are now different in the two equations, whereas in the open loop case we had the same term, $\hat{v}^i(t)$. Nevertheless, since $\hat{v}^i(x)$ is Lipschitz $|\hat{v}^i(\hat{y}^i(t)) - \hat{v}^i(\hat{x}^i(t))|$ is estimated by $|\hat{y}^i(t) - \hat{x}^i(t)|$ and thus the reasoning of the open loop case carries over, with additional terms. We obtain that $\hat{x}^i(t)$ approximates $\hat{y}^i(t)$. Since we consider only Lipschitz local feedbacks, the reasoning of the open loop case to show the property of approximate Nash equilibrium will apply again and the result can then be obtained.

Chapter 6
Linear Quadratic Models

6.1 Setting of the Model

The linear quadratic model has been developed in [10]. See also [2, 20, 22]. We highlight here the results. We take

$$f(x,m,v) = \frac{1}{2}\left[x^*Qx + v^*Rv + \left(x - S\int \xi m(\xi)d\xi\right)^* \bar{Q}\left(x - S\int \xi m(\xi)d\xi\right)\right]$$
(6.1)

$$g(x,m,v) = Ax + \bar{A}\int \xi m(\xi)d\xi + Bv$$
(6.2)

$$h(x,m) = \frac{1}{2}\left[x^*Q_Tx + \left(x - S_T\int \xi m(\xi)d\xi\right)^* \bar{Q}_T\left(x - S_T\int \xi m(\xi)d\xi\right)\right].$$
(6.3)

We also take $\sigma(x) = \sigma$, constant and set $a = \frac{1}{2}\sigma\sigma*$. Finally we assume that $m_0(x)$ is a gaussian with mean $\bar{x}_0$ and variance Γ_0. Of course, the matrix A is not the operator $-\text{tr}aD^2$.

6.2 Solution of the Mean Field Game Problem

We need to solve the system of HJB-FP equations (3.11), which reads

$$-\frac{\partial u}{\partial t} - \text{tr }aD^2u = -\frac{1}{2}Du^*BR^{-1}B^*Du + Du \cdot \left(Ax + \bar{A}\int \xi m(\xi)d\xi\right)$$

$$+ \frac{1}{2}\left[x^*Qx + \left(x - S\int \xi m(\xi)d\xi\right)^* \bar{Q}\left(x - S\int \xi m(\xi)d\xi\right)\right]$$

A. Bensoussan et al., *Mean Field Games and Mean Field Type Control Theory*, SpringerBriefs in Mathematics, DOI 10.1007/978-1-4614-8508-7_6, © Alain Bensoussan, Jens Frehse, Phillip Yam 2013

$$u(x,T) = \frac{1}{2}\left[x^*Q_Tx + \left(x - S_T\int \xi m(\xi)d\xi\right)^*\bar{Q}_T\left(x - S_T\int \xi m(\xi)d\xi\right)\right]$$

$$(6.4)$$

$$\frac{\partial m}{\partial t} - \text{tr } aD^2m + \text{div}\left(m\left(Ax + \bar{A}\int \xi m(\xi)d\xi - BR^{-1}B^*Du\right)\right) = 0$$

$$m(x,0) = m_0(x). \tag{6.5}$$

We look for a solution $u(x,t)$ of the form

$$u(x,t) = \frac{1}{2}x^*P(t)x + x^*r(t) + s(t) \tag{6.6}$$

then

$$Du = P(t)x + r(t), \quad D^2u = P(t).$$

Equation (6.5) becomes

$$\frac{\partial m}{\partial t} - \text{tr } aD^2m + \text{div}\left(m\left((A - BR^{-1}B^*P)x - BR^{-1}B^*r + \bar{A}\int \xi m(\xi)d\xi\right)\right) = 0$$

Setting $z(t) = \int xm(x,t)dx$, it follows easily that

$$\frac{dz}{dt} = (A + \bar{A} - BR^{-1}B^*P(t))z(t) - BR^{-1}B^*r(t) \tag{6.7}$$

$$z(0) = \bar{x}_0. \tag{6.8}$$

Going back to the HJB equation and using the expression (6.6), we obtain that $P(t)$ is the solution of the Riccati equation

$$\frac{dP}{dt} + PA + A^*P - PBR^{-1}B^*P + Q + \bar{Q} = 0$$

$$P(T) = Q_T + \bar{Q}_T. \tag{6.9}$$

This Riccati equation has a unique positive symmetric solution. To get $r(t)$ we need to solve the coupled system of differential equations

$$\frac{dz}{dt} = (A + \bar{A} - BR^{-1}B^*P(t))z(t) - BR^{-1}B^*r(t) \tag{6.10}$$

$$z(0) = \bar{x}_0 \tag{6.11}$$

$$-\frac{dr}{dt} = (A^* - P(t)BR^{-1}B^*)r(t) + (P(t)\bar{A} - Q\bar{S})z(t) \tag{6.12}$$

$$r(T) = -\bar{Q}_T S_T z(T). \tag{6.13}$$

It is easy to get $s(t)$ by the formula

$$s(t) = \frac{1}{2}z(T)^* S_T^* \bar{Q}_T S_T z(T)$$

$$+ \int_t^T \left[\operatorname{tr} aP(s) - \frac{1}{2}r(s)^* BR^{-1}B^* r(s) \right.$$

$$\left. + r(s)^* \bar{A}z(s) + \frac{1}{2}z(s)^* S^* \bar{Q}S z(s) \right] ds. \tag{6.14}$$

So the solvability condition reduces to solving the system (6.10) and (6.12). The function $u(x,t)$ is given by (6.6). The function $m(x,t)$ can then be obtained easily and it is a gaussian. However, (6.10) and (6.12) do not necessarily have a solution. Conditions are needed. We will come back to this point after expressing the maximum principle.

We write the maximum principle, applying (3.14) and (3.15). We obtain stocastic processes $X(t), V(t), Y(t), Z(t)$ which must satisfy

$$dX = (AX - BR^{-1}B^* Y + \bar{A}EX(t))dt + \sigma dw$$

$$-dY = (A*Y + (Q + \bar{Q})X - \bar{Q}S EX(t))dt - Zdw$$

$$X(0) = x_0$$

$$Y(T) = (Q_T + \bar{Q}_T)X(T) - \bar{Q}_T S_T EX(T) \tag{6.15}$$

$$V(t) = -R^{-1}B^* Y(t). \tag{6.16}$$

It follows that $\bar{X}(t) = EX(t)$, $\bar{Y}(t) = EY(t)$ must satisfy the system

$$\frac{d\bar{X}}{dt} = (A + \bar{A})\bar{X} - BR^{-1}B^* \bar{Y}$$

$$\bar{X}(0) = \bar{x}_0$$

$$-\frac{d\bar{Y}}{dt} = A*\bar{Y} + (Q + \bar{Q}(I - S))\bar{X}$$

$$\bar{Y}(T) = (Q_T + \bar{Q}_T(I - S_T))\bar{X}(T). \tag{6.17}$$

If we can solve this system, then (6.15) can be readily solved and the optimal control is obtained. Now (6.17) is identical to (6.10) and (6.12) with the correspondence

$$z(t) = \bar{X}(t) \tag{6.18}$$

$$r(t) = \bar{Y}(t) - P(t)\bar{X}(t). \tag{6.19}$$

Even though (6.17) is equivalent to (6.10) and (6.12), the situation differs when one tries to define sufficient conditions for a solution of (6.10) and (6.12) or for a solution of (6.17) to exist. Indeed, (6.10) and (6.12) involve $P(t)$, so the condition will be expressed in terms of $P(t)$ and thus is not easily checkable. The system (6.17) involves the data directly. Thus, simpler conditions of a solution can be obtained (see [10]) for a complete discussion. Moreover, we can see that (6.17) is related to a nonsymmetric Riccati equation. Indeed, we have

$$\bar{Y}(t) = \Sigma(t)\bar{X}(t) \tag{6.20}$$

with

$$\frac{d\Sigma}{dt} + \Sigma(A + \bar{A}) + A^*\Sigma - \Sigma BR^{-1}B^*\Sigma + Q + \bar{Q}(I - S) = 0$$

$$\Sigma(T) = Q_T + \bar{Q}_T(I - S_T). \tag{6.21}$$

It follows that

$$r(t) = (\Sigma(t) - P(t))\bar{X}(t). \tag{6.22}$$

However $\Sigma(t) - P(t)$ is not solution of a single equation. Equation (6.21) is not standard since it not symmetric. The existence of a solution of (6.17) is equivalent to finding a solution of (6.21).

6.3   Solution of the Mean Field Type Problem

We apply (4.12). We first need to compute

$$\frac{\partial H}{\partial m}(\xi, m, q)(x) = q^*\bar{A}x - \left(\xi - S\int \eta m(\eta)d\eta\right)^* \bar{Q}Sx \tag{6.23}$$

hence

$$\int \frac{\partial H}{\partial m}(\xi, m, Du(\xi))(x)m(\xi)d\xi$$

$$= \left(\int Du(\xi)m(\xi)d\xi\right)^* \bar{A}x - \left(\int \eta m(\eta)d\eta\right)^* (I - S)^*\bar{Q}Sx. \tag{6.24}$$

The HJB equation reads

$$-\frac{\partial u}{\partial t} - \operatorname{tr} aD^2 u = -\frac{1}{2} Du^* BR^{-1} B^* Du + Du. \left(Ax + \bar{A} \int \xi m(\xi) d\xi \right)$$

$$+ \frac{1}{2} \left[x^* Qx + \left(x - S \int \xi m(\xi) d\xi \right)^* \bar{Q} \left(x - S \int \xi m(\xi) d\xi \right) \right]$$

$$+ \left(\int Du(\xi) m(\xi) d\xi \right)^* \bar{A} x - \left(\int \xi m(\xi) d\xi \right)^* (I - S)^* \bar{Q} Sx$$

$$\tag{6.25}$$

$$u(x, T) = \frac{1}{2} \left[x^* Q_T x + \left(x - S_T \int \xi m(\xi) d\xi \right)^* \bar{Q}_T \left(x - S_T \int \xi m(\xi) d\xi \right) \right]$$

$$- \left(\int \xi m(\xi) d\xi \right)^* (I - S_T)^* \bar{Q}_T S_T x. \tag{6.26}$$

For the FP equation, there is no change. It is

$$\frac{\partial m}{\partial t} - \operatorname{tr} aD^2 m + \operatorname{div} \left(m \left(Ax + \bar{A} \int \xi m(\xi) d\xi - BR^{-1} B^* Du \right) \right) = 0$$

$$m(x, 0) = m_0(x). \tag{6.27}$$

We look for a solution

$$u(x, t) = \frac{1}{2} x^* P(t) x + \rho^*(t) + \tau(t). \tag{6.28}$$

Calling again

$$z(t) = \int x m(x, t) dx,$$

we get the differential equation

$$\frac{dz}{dt} = (A + \bar{A} - BR^{-1} B^* P(t)) z(t) - BR^{-1} B^* \rho(t)$$

$$z(0) = \bar{x}_0.$$

Replacing $u(x, t)$ in the HJB equation (6.25) and identifying terms we obtain

$$\frac{dP}{dt} + PA + A^* P - PBR^{-1} B^* P + Q + \bar{Q} = 0$$

$$P(T) = Q_T + \bar{Q}_T \tag{6.29}$$

and the pair $z(t), \rho(t)$ must be a solution of the system

$$\frac{dz}{dt} = (A + \bar{A} - BR^{-1}B^*P(t))z(t) - BR^{-1}B^*\rho(t)$$

$$z(0) = \bar{x}_0 \tag{6.30}$$

$$-\frac{d\rho}{dt} = (A^* + \bar{A}^* - P(t)BR^{-1}B^*)\rho(t)$$

$$+ (P(t)\bar{A} + \bar{A}^*P - \bar{Q}S - S^*\bar{Q} + S^*\bar{Q}S)z(t)$$

$$\rho(T) = (-\bar{Q}_T S_T - S_T^*\bar{Q}_T + S_T^*\bar{Q}_T S_T)z(T). \tag{6.31}$$

We note that $P(t)$ is identical to the case of a mean field game; see (6.9). The system $z(t), \rho(t)$ is different from (6.10) and (6.12). We will see in considering the stochastic maximum principle that it always has a solution.

To write the stochastic maximum principle, we use (4.19). We obtain stochastic processes $X(t), V(t), Y(t), and Z(t)$, which must satisfy

$$dX = (AX - BR^{-1}B^*Y + \bar{A}EX(t))dt + \sigma dw$$

$$-dY = (A*Y + (Q + \bar{Q})X + \bar{A}^*EY(t) - (\bar{Q}S - S^*\bar{Q}(I - S))EX(t))dt - Zdw \tag{6.32}$$

$$X(0) = x_0$$

$$Y(T) = (Q_T + \bar{Q}_T)X(T) - (\bar{Q}_T S_T - S_T^*\bar{Q}_T(I - S_T))EX(T)$$

$$V(t) = -R^{-1}B^*Y(t). \tag{6.33}$$

Writing $\bar{X}(t) = EX(t)$, $\bar{Y}(t) = EY(t)$ we deduce the system

$$\frac{d\bar{X}}{dt} = (A + \bar{A})\bar{X} - BR^{-1}B^*\bar{Y}$$

$$\bar{X}(0) = \bar{x}_0$$

$$-\frac{d\bar{Y}}{dt} = (A + \bar{A})*\bar{Y} + (Q + (I - S)^*\bar{Q}(I - S))\bar{X}$$

$$\bar{Y}(T) = (Q_T + (I - S_T)^*\bar{Q}_T(I - S_T))\bar{X}(T). \tag{6.34}$$

Conversely, in the case of mean field games [see (6.17)], this system always has a solution, provided that we assume

$$Q + (I - S)^*\bar{Q}(I - S) \geq 0, \ Q_T + (I - S_T)^*\bar{Q}_T(I - S_T) \geq 0. \tag{6.35}$$

In particular, if we write $\bar{Y}(t) = \Sigma(t)\bar{X}(t)$, then $\Sigma(t)$ is solution of the symmetric Riccati equation

$$\frac{d\Sigma}{dt} + \Sigma(A+\bar{A}) + (A+\bar{A})^*\Sigma - \Sigma BR^{-1}B^*\Sigma + Q + (I-S)^*\bar{Q}(I-S) = 0$$

$$\Sigma(T) = Q_T + (I-S_T)^*\bar{Q}_T(I-S_T). \qquad (6.36)$$

So we can see in the linear quadratic case that the mean field type control problem has solutions more generally than the mean field game problem.

6.4 The Mean Variance Problem

The mean variance problem is the extension in continuous time for a finite horizon of the Markowitz optimal portfolio theory. Without referring to the background of the problem, it can be stated as follows, mathematically. The state equation is

$$dx = rxdt + xv \cdot (\alpha dt + \sigma dw)$$

$$x(0) = x_0 \qquad (6.37)$$

$x(t)$ is scalar, r is a positive constant, α is a vector in $\mathbb{R}^m$, and σ is a matrix in $\mathcal{L}(\mathbb{R}^d; \mathbb{R}^m)$. All can depend on time and they are deterministic quantities. $v(t)$ is the control in $\mathbb{R}^m$. We note that, conversely to our general framework, the control affects the volatility term. The objective function is

$$J(v(.)) = Ex(T) - \frac{\gamma}{2}\text{var}(x(T)) \qquad (6.38)$$

which we want to maximize.

Because of the variance term, the problem is not a standard stochastic control problem. It is a mean field type control problem, since one can write

$$J(v(.)) = E(x(T) - \frac{\gamma}{2}x(T)^2) + \frac{\gamma}{2}(Ex(T))^2. \qquad (6.39)$$

Because of the presence of the control in the volatility term, we need to reformulate our general approach of mean field type problems, which we shall do formally, and without details. We consider a feedback control $v(x,s)$ and the corresponding state $x_{v(.)}(t)$ solution of (6.37) when the control is replaced by the feedback. We associate the probability density $m_{v(.)}(x,t)$ solution of

$$\frac{\partial m_{v(.)}}{\partial t} + \frac{\partial}{\partial x}(xm_{v(.)}(r + \alpha \cdot v(x))) - \frac{1}{2}\frac{\partial^2}{\partial x^2}(x^2 m_{v(.)}|\sigma^* v(x)|^2) = 0$$

$$m_{v(.)}(x,0) = \delta(x - x_0). \qquad (6.40)$$

The functional (6.39) can be written as

$$J(v(.)) = \int m_{v(.)}(x,T)(x - \frac{\gamma}{2}x^2)dx + \frac{\gamma}{2}\left(\int m_{v(.)}(x,T)xdx\right)^2. \qquad (6.41)$$

Let $\hat{v}(x,t)$ be an optimal feedback, and $m(t) = m_{\hat{v}(.)}(t)$. We compute the Frechet derivative

$$\tilde{m}(x,t) = \frac{d}{d\theta}(m_{\hat{v}(.)+\theta v(.)})_{|\theta=0}$$

which is the solution of

$$\frac{\partial \tilde{m}}{\partial t} + \frac{\partial}{\partial x}(x\tilde{m}(r + \alpha \cdot \hat{v}(x))) - \frac{1}{2}\frac{\partial^2}{\partial x^2}(x^2\tilde{m}|\hat{v}^*(x)\sigma|^2) = -\frac{\partial}{\partial x}(xm\alpha.v)$$

$$+ \frac{\partial^2}{\partial x^2}(x^2m\hat{v}^*(x)\sigma \cdot v^*\sigma)$$

$$\tilde{m}(x,0) = 0. \qquad (6.42)$$

We deduce the Gateux differential of the objective functional

$$\frac{d}{d\theta}J(\hat{v}(.)+\theta v(.))_{|\theta=0} = \int \tilde{m}(x,T)\left(x - \frac{\gamma}{2}x^2\right)dx + \gamma\int m(x,T)xdx\int \tilde{m}(x,T)xdx. \qquad (6.43)$$

We next introduce the function $u(x,t)$ solution of

$$-\frac{\partial u}{\partial t} - x(r + \alpha \cdot \hat{v}(x))\frac{\partial u}{\partial x} - \frac{1}{2}x^2|\sigma\hat{v}^*(x)|^2\frac{\partial^2 u}{\partial x^2} = 0$$

$$u(x,T) = x - \frac{\gamma}{2}x^2 + \gamma x\int m(\xi,T)\xi d\xi \qquad (6.44)$$

then

$$\frac{d}{d\theta}J(\hat{v}(.)+\theta v(.))_{|\theta=0} = \int \tilde{m}(x,T)u(x,T)dx$$

$$= \int_0^T\int_{\mathbb{R}} m(x,t)v^*(x,t)\left(x\frac{\partial u}{\partial x}\alpha + x^2\frac{\partial^2 u}{\partial x^2}\sigma\sigma^*\hat{v}(x,t)\right)dxdt.$$

Since $\hat{v}(.)$ maximizes $J(v(.))$, we obtain

$$v^*(x,t)\left(x\frac{\partial u}{\partial x}\alpha + x^2\frac{\partial^2 u}{\partial x^2}\sigma\sigma^*\hat{v}(x,t)\right) \leq 0, \text{ a.e. } x,t, \forall v(x,t)$$

hence

$$\hat{v}(x,t) = -\frac{\dfrac{\partial u}{\partial x}}{x\dfrac{\partial^2 u}{\partial x^2}}(\sigma\sigma^*)^{-1}\alpha. \tag{6.45}$$

So the pair $u(x,t), m(x,t)$ satisfies

$$-\frac{\partial u}{\partial t} - xr\frac{\partial u}{\partial x} + \frac{1}{2}\frac{\left(\dfrac{\partial u}{\partial x}\right)^2}{\dfrac{\partial^2 u}{\partial x^2}}\alpha^*(\sigma\sigma^*)^{-1}\alpha = 0$$

$$u(x,T) = x - \frac{\gamma}{2}x^2 + \gamma x \int m(\xi,T)\xi d\xi \tag{6.46}$$

$$\frac{\partial m}{\partial t} + r\frac{\partial(xm)}{\partial x} - \frac{\partial}{\partial x}\left(m\frac{\dfrac{\partial u}{\partial x}}{\dfrac{\partial^2 u}{\partial x^2}}\right)\alpha^*(\sigma\sigma^*)^{-1}\alpha$$

$$-\frac{1}{2}\frac{\partial^2}{\partial x^2}\left(m\frac{(\dfrac{\partial u}{\partial x})^2}{(\dfrac{\partial^2 u}{\partial x^2})^2}\right)\alpha^*(\sigma\sigma^*)^{-1}\alpha - 0$$

$$m(x,0) = \delta(x - x_0). \tag{6.47}$$

We can solve this system explicitly. We look for

$$u(x,t) = -\frac{1}{2}P(t)x^2 + s(t)x + \rho(t). \tag{6.48}$$

We also define

$$q(t) = \int m(\xi,t)\xi d\xi. \tag{6.49}$$

From (6.47) we obtain easily

$$\frac{1}{2}\dot{P} + \left(r - \frac{1}{2}\alpha^*(\sigma\sigma^*)^{-1}\alpha\right)P = 0$$

$$P(T) = \gamma$$

$$\dot{s} + (r - \alpha^*(\sigma\sigma^*)^{-1}\alpha)s = 0$$

$$s(T) = 1 + \gamma q(T)$$

$$\dot{\rho} + \frac{1}{2}\frac{s^2}{P}\alpha^*(\sigma\sigma^*)^{-1}\alpha = 0$$

$$\rho(T) = 0.$$

We obtain

$$P(t) = \gamma \exp\left(\int_t^T (2r - \alpha^*(\sigma\sigma^*)^{-1}\alpha)d\tau\right)$$

$$s(t) = (1 + \gamma q(T)) \exp\left(\int_t^T (r - \alpha^*(\sigma\sigma^*)^{-1}\alpha)d\tau\right)$$

$$\rho(t) = \int_t^T \frac{1}{2}\frac{s^2}{P}\alpha^*(\sigma\sigma^*)^{-1}\alpha(\tau)d\tau. \qquad (6.50)$$

We need to fix $q(T)$. Equation (6.47) becomes

$$\frac{\partial m}{\partial t} + r\frac{\partial(xm)}{\partial x} - \frac{\partial}{\partial x}\left(m\left(x - \frac{s}{P}\right)\right)\alpha^*(\sigma\sigma^*)^{-1}\alpha$$

$$-\frac{1}{2}\frac{\partial^2}{\partial x^2}\left(m\left(x - \frac{s}{P}\right)^2\right)\alpha^*(\sigma\sigma^*)^{-1}\alpha = 0$$

$$m(x,0) = \delta(x - x_0). \qquad (6.51)$$

If we test this equation with x we obtain easily

$$\dot{q} - (r - \alpha^*(\sigma\sigma^*)^{-1}\alpha))q = \frac{s}{P}\alpha^*(\sigma\sigma^*)^{-1}\alpha$$

$$= \frac{1 + \gamma q(T)}{\gamma}\alpha^*(\sigma\sigma^*)^{-1}\alpha \exp\left(-\int_t^T rd\tau\right)$$

$$q(0) = x_0.$$

We deduce easily

$$q(T) = x_0 \exp\left(\int_0^T (r - \alpha^*(\sigma\sigma^*)^{-1}\alpha)d\tau\right)$$

$$+ L\frac{1 + \gamma q(T)}{\gamma}\left[1 - \exp\left(-\int_0^T \alpha^*(\sigma\sigma^*)^{-1}\alpha)d\tau\right)\right]$$

and we obtain

$$q(T) = x_0 \exp\left(\int_0^T rd\tau\right) + \frac{1}{\gamma}\left[\exp\left(\int_0^T \alpha^*(\sigma\sigma^*)^{-1}\alpha)d\tau\right) - 1\right]. \qquad (6.52)$$

This completes the definition of the function $u(x,t)$. The optimal feedback is defined by [see (6.45)]

$$\hat{v}(x,t) = -(\sigma\sigma^*)^{-1}\alpha + \frac{1}{x}\frac{1+\gamma q(T)}{\gamma}\exp\left(-\int_t^T rd\tau\right). \qquad (6.53)$$

We see that this optimal feedback depends on the initial condition x_0.

If we take the time consistency approach, we consider the family of problems

$$dx = rxds + xv(x,s).(\alpha dt + \sigma dw), \, s > t$$
$$x(t) = x \qquad\qquad\qquad\qquad\qquad (6.54)$$

and the payoff

$$J_{x,t}(v(.)) = E\left(x(T) - \frac{\gamma}{2}x(T)^2\right) + \frac{\gamma}{2}(Ex(T))^2. \qquad (6.55)$$

Denote by $\hat{v}(x,s)$ an optimal feedback and set $V(x,t) = J_{x,t}(\hat{v}(.))$. We define

$$\Psi(x,t;T) = E\hat{x}_{xt}(T)$$

where $\hat{x}_{xt}(s)$ is the solution of (6.54) for the optimal feedback. The function $\Psi(x,t;T)$ is the solution of

$$\frac{\partial\Psi}{\partial t} + \frac{\partial\Psi}{\partial x}(rx + x\hat{v}(x,t)^*\alpha) + \frac{1}{2}x^2\frac{\partial^2\Psi}{\partial x^2}|\sigma^*\hat{v}(x,t)|^2 = 0$$
$$\Psi(x,T;T) = x.$$

We can write

$$V(x,t) = E\left(\hat{x}_{xt}(T) - \frac{\gamma}{2}\hat{x}_{xt}(T)^2\right) + \frac{\gamma}{2}(\Psi(x,t;T))^2.$$

We consider the spike modification

$$\bar{v}(x,s) = \begin{cases} v & t < s < t+\varepsilon \\ \hat{v}(x,s) & s > t+\varepsilon \end{cases}$$

then

$$J_{x,t}(\bar{v}(.)) = E(\hat{x}_{x(t+\varepsilon),t+\varepsilon}(T) - \frac{\gamma}{2}\hat{x}_{x(t+\varepsilon),t+\varepsilon}(T)^2)$$
$$+ \frac{\gamma}{2}(E\Psi(x(t+\varepsilon),t+\varepsilon;T))^2$$

where $x(t+\varepsilon)$ corresponds to the solution of (6.54) at time $t+\varepsilon$ for the feedback equal to the constant v. We note that

$$EV(x(t+\varepsilon),t+\varepsilon) = E(\hat{x}_{x(t+\varepsilon),t+\varepsilon}(T) - \frac{\gamma}{2}\hat{x}_{x(t+\varepsilon),t+\varepsilon}(T)^2)$$
$$+ \frac{\gamma}{2}E(\Psi(x(t+\varepsilon),t+\varepsilon;T))^2$$

so we need to compare $(E\Psi(x(t+\varepsilon),t+\varepsilon;T))^2$ with $E(\Psi(x(t+\varepsilon),t+\varepsilon;T))^2$. Using the techniques of Sect. 4.4 we see easily that

$$(E\Psi(x(t+\varepsilon),t+\varepsilon;T))^2 - E(\Psi(x(t+\varepsilon),t+\varepsilon;T))^2$$
$$= -\varepsilon x^2 \frac{\partial^2 \Psi}{\partial x^2}(x,t;T)|\sigma^* v|^{2'} + 0(\varepsilon).$$

Therefore,

$$V(x,t) \geq J_{x,t}(\bar{v}(.)) = EV(x(t+\varepsilon),t+\varepsilon)$$
$$- \frac{\varepsilon}{2}x^2 \frac{\partial^2 \Psi}{\partial x^2}(x,t;T)|\sigma^* v|^2 + 0(\varepsilon).$$

Expanding $EV(x(t+\varepsilon),t+\varepsilon)$ we obtain the HJB equation

$$\frac{\partial V}{\partial t} + \frac{\partial V}{\partial x}rx + \max_v[x\frac{\partial V}{\partial x}v^*\alpha + \frac{x^2}{2}(\frac{\partial^2 V}{\partial x^2} - \gamma\frac{\partial^2 \Psi}{\partial x^2}(x,t;T))v^*\sigma\sigma^* v] = 0$$
$$V(x,T) = x. \qquad (6.56)$$

A direct checking shows that

$$V(x,t) = x\exp(r(T-t)) + \frac{1}{2\gamma}\int_t^T \alpha^*(\sigma\sigma^*)\alpha ds \qquad (6.57)$$

$$\Psi(x,t) = x\exp(r(T-t)) + \frac{1}{\gamma}\int_t^T \alpha^*(\sigma\sigma^*)\alpha ds$$

and

$$\hat{v}(x,t) = \frac{\exp(-r(T-t))}{x\gamma}(\sigma\sigma^*)\alpha \qquad (6.58)$$

This optimal control satisfies the time consistency property (it does not depend on the initial condition).

6.5 Approximate *N* Player Differential Game

From the definition of $f(x,m,v)$, $g(x,m,v)$, $and h(x,m)$—see (6.1), (6.2), and (6.3)—the differential game (5.1) and (5.2) becomes

$$dx^i = \left(Ax^i + Bv^i + \bar{A}\frac{1}{N-1}\sum_{\substack{j=1\neq i}}^{N} x^j \right) dt + \sigma dw^i \qquad (6.59)$$

$$x^i(0) = x_0^i \qquad (6.60)$$

$$\mathcal{J}^i(v(.)) = \frac{1}{2}E\left[\int_0^T (x^i)^* Q x^i \right.$$

$$+ \left(x^i - S\frac{1}{N-1}\sum_{\substack{j=1\neq i}}^{N} x^j \right)^* \bar{Q}\left(x^i - S\frac{1}{N-1}\sum_{\substack{j=1\neq i}}^{N} x^j \right) + (v^i)^* R v^i \bigg] dt$$

$$\left. + \frac{1}{2}E\left[(x^i)^* Q_T\, x^i(T) + \left(x^i - S_T\frac{1}{N-1}\sum_{\substack{j=1\neq i}}^{N} x^j \right)^* \bar{Q}_T\left(x^i - S_T\frac{1}{N-1}\sum_{\substack{j=1\neq i}}^{N} x^j \right)(T) \right].$$

$$(6.61)$$

The reasoning developed in Sect. 5.4, can be applied in the present situation. Considering the optimal feedback

$$\hat{v}(x) = -R^{-1}B^*(P(t)x + r(t)) \qquad (6.62)$$

where $P(t)$ and $r(t)$ have been defined in (6.9) and (6.12). We consider the stochastic process $\hat{v}(t) = \hat{v}(\hat{x}(t))$ where $\hat{x}(t)$ is the optimal trajectory for the control problem with the mean field term

$$d\hat{x} = ((A - BR^{-1}B^*P(t))\hat{x} - R^{-1}B^*r(t) + \bar{A}E\hat{x}(t))dt + \sigma dw$$
$$\hat{x}(0) = x_0 \qquad (6.63)$$

so

$$\hat{v}(t) = -R^{-1}B^*(P(t)\hat{x}(t) + r(t)). \qquad (6.64)$$

We then consider *N* independent Wiener processes $w^i(.)$ and initial conditions x_0^i. We duplicate the stochastic process $\hat{v}(t)$ into $\hat{v}^i(t)$. These processes constitute an approximate open-loop Nash equilibrium for the objective functionals (6.61). We can also consider the feedback (6.62) and duplicate for all players. We will also obtain an approximate Nash equilibrium among smooth feedbacks (see Sect. 5.5).

Chapter 7
Stationary Problems

7.1 Preliminaries

We shall consider only mean field games, but mean field type control can also
be considered. To obtain stationary problems, Lasry and Lions [27] consider
ergodic situations. This introduces an additional difficulty. It is, however, possible to
motivate stationary problems that correspond to infinite horizon discounted control
problems. The price to pay concerns the N player differential game associated with
the mean field game. It is less natural than the one used in the time-dependent case.
However, other interpretations are possible, which do not lead to the same difficulty.

The natural counterpart of (3.11) is

$$Au + \alpha u = H(x, m, Du)$$

$$A^* m + \text{div } (G(x, m, Du)m) + \alpha m = \alpha m_0, \qquad (7.1)$$

where α is a positive number (the discount). Note that when m_0 is a probability,
the solution m is also a probability. We shall simplify a little bit with respect to the
time-dependent case. We consider $f(x, m, v)$ as in Sect. 2.1, but we take $g(x, v)$ as
not being dependent on m. Then, considering the Lagrangian

$$f(x, m, v) + q.g(x, v)$$

we suppose that the minimum is attained at a point $\hat{v}(x, m, q)$, and this function is
well-defined and sufficiently smooth. We next define the Hamiltonian

$$H(x, m, q) = \inf_v [f(x, m, v) + q.g(x, v)] \qquad (7.2)$$

$$= f(x, m, \hat{v}(x, m, q)) + q.g(x, \hat{v}(x, m, q)) \qquad (7.3)$$

A. Bensoussan et al., *Mean Field Games and Mean Field Type Control Theory*,
SpringerBriefs in Mathematics, DOI 10.1007/978-1-4614-8508-7_7,
© Alain Bensoussan, Jens Frehse, Phillip Yam 2013

and

$$G(x,m,q) = g(x,\hat{v}(x,m,q)). \quad (7.4)$$

These are the functions entering into (7.1).

7.2   Mean Field Game Set-Up

Consider a feedback $v(.)$ and a probability density $m(.)$. We construct the state equation associated to the feedback control

$$dx = g(x(t), v(x(t)))dt + \sigma(x(t))dw(t)$$

$$x(0) = x_0. \quad (7.5)$$

We then define the cost functional

$$J(v(.), m(.)) = E \int_0^{+\infty} \exp(-\alpha t) f(x(t), m, v(x(t)))dt \quad (7.6)$$

and denote by $p_{v(.)}(x,t)$ the probability distribution of the process $x(t)$. We use $p_{v(.)}(x,t)$ instead of $m_{v(.)}(x,t)$ to avoid confusion. A pair $\hat{v}(.)$ and $m(.)$ is a solution of the mean field game problem, whenever

$$J(\hat{v}(.), m(.)) \le J(v(.), m(.)), \ \forall v(.) \quad (7.7)$$

$$m = \alpha \int_0^{+\infty} \exp(-\alpha t) p_{\hat{v}(.)}(t)dt. \quad (7.8)$$

From standard control theory, it is clear that

$$J(\hat{v}(.), m(.)) = \int_{\mathbb{R}^n} m_0(x)u(x)dx \quad (7.9)$$

in which $u(x)$ is the solution of the first equation (7.1), and

$$\hat{v}(x) = \hat{v}(x, m, Du(x)) \quad (7.10)$$

so $G(x, m, Du) = g(x, \hat{v}(x))$. Now $p_{\hat{v}(.)}(x,t)$ is the solution of

$$\frac{\partial p_{\hat{v}(.)}}{\partial t} + A^* p_{\hat{v}(.)} + \operatorname{div}\,(g(x, \hat{v}(x))p_{\hat{v}(.)}) = 0$$

$$p_{\hat{v}(.)}(x,0) = m_0(x). \quad (7.11)$$

If we compute $\alpha \int_0^{+\infty} \exp(-\alpha t) \, p_{\hat{v}(.)}(t) dt$ we easily see that it satisfies the second equation (7.1). Note that

$$\alpha J(\hat{v}(.), m(.)) = \int_{\mathbb{R}^n} m(x) f(x, m, \hat{v}(x)) dx. \qquad (7.12)$$

For any feedback $v(.)$ we can introduce $p_{v(.)}(x,t)$ as the solution of

$$\frac{\partial p_{v(.)}}{\partial t} + A^* p_{v(.)} + \mathrm{div}\,(g(x, v(x)) p_{v(.)}) = 0$$

$$p_{v(.)}(x,0) = m_0(x)$$

and let

$$p_{v(.),\alpha}(x) = \alpha \int_0^{+\infty} \exp(-\alpha t) \, p_{v(.)}(x,t) dt$$

then we can write

$$\alpha J(v(.), m(.)) = \int_{\mathbb{R}^n} p_{v(.),\alpha}(x) f(x, m, v(x)) dx \qquad (7.13)$$

so we have the fixed-point property $m = p_{\hat{v}(.),\alpha}(x)$, with $\hat{v}(.)$ depending on m from formula (7.10).

7.3 Additional Interpretations

We restrict the model to

$$f(x, m, v) = f(x, v) + \frac{\partial \Phi(m)}{\partial m}(x), \qquad (7.14)$$

where $\Phi(m)$ is a functional on $L^1(\mathbb{R}^n)$. For any feedback $v(.)$ consider the probability $p_{v(.),\alpha}(x)$. It is the solution of

$$A^* p_{v(.),\alpha} + \mathrm{div}\,(g(x, v(x)) p_{v(.),\alpha}) + \alpha p_{v(.),\alpha} = \alpha m_0. \qquad (7.15)$$

We define an objective functional

$$\alpha J(v(.)) = \int_{\mathbb{R}^n} f(x, v(x)) p_{v(.),\alpha}(x) dx + \Phi(p_{v(.),\alpha}). \qquad (7.16)$$

We can obtain the Gâteaux differential

$$\alpha \frac{d J(v(.) + \theta \tilde{v}(.))}{d\theta}\Big|_{\theta=0} = \int_{\mathbb{R}^n} \frac{\partial L}{\partial v}(x, v(x), Du_{v(.)}(x)) p_{v(.),\alpha}(x) \tilde{v}(.) dx, \qquad (7.17)$$

where $u_{v(.)}(x)$ is the solution of

$$Au_{v(.)} + \alpha u_{v(.)} = f(x,v(x)) + g(x,v(x)).Du_{v(.)} + \frac{\partial\Phi(p_{v(.),\alpha})}{\partial m}(x) \qquad (7.18)$$

and

$$L(x,v,q) = f(x,v) + q.g(x,v).$$

A necessary condition of optimality is

$$\frac{\partial L}{\partial v}(x,v(x),Du_{v(.)}(x)) = 0 \qquad (7.19)$$

which, if convexity in v holds, implies that the optimal feedback minimizes in v, the Lagrangian $L(x,v,Du(x))$. We can define the Hamiltonian

$$H(x,q) = \inf_{v}(f(x,v) + q.g(x,v)) \qquad (7.20)$$

and

$$H(x,m,q) = H(x,q) + \frac{\partial\Phi(m)}{\partial m}(x). \qquad (7.21)$$

Considering the point of minimum $\hat{v}(x,q)$ and

$$G(x,m,q) = G(x,q) = g(x,\hat{v}(x,q)) \qquad (7.22)$$

then the system (7.1) can be interpreted as a necessary condition of optimality for the problem of minimizing $J(v(.))$.

Still, in the convex case, we can also interpret (7.1) as a necessary condition of optimality for a control problem of the HJB equation.

Define next the conjugate of $\Phi(m)$ by

$$\Phi^*(z(.)) = \sup_{m}\left[\Phi(m) - \int_{\mathbb{R}^n} z(x)m(x)dx\right] \qquad (7.23)$$

For any $z(.) \in L^\infty(\mathbb{R}^n)$, define $m_{z(.)}(x)$ to be the point of supremum in (7.23). It satisfies

$$\frac{\partial\Phi(m_{z(.)})}{\partial m}(x) = z(x). \qquad (7.24)$$

Define next $u_{z(.)}(x)$ to be the solution of

$$Au_{z(.)} + \alpha u_{z(.)} = H(x,Du_{z(.)}) + z(x). \qquad (7.25)$$

In this equation $z(x)$ appears as a control, and the corresponding state is $u_{z(.)}$. The objective functional is defined by

$$\mathcal{K}(z(.)) = \Phi^*(z(.)) + \alpha \int_{\mathbb{R}^n} m_0(x) u_{z(.)}(x) dx. \qquad (7.26)$$

We can look for a necessary condition of optimality in minimizing $\mathcal{K}(z(.))$. We first have

$$\frac{d}{d\theta} \Phi^*(z(.) + \theta \tilde{z}(.))|_{\theta=0} = -\int_{\mathbb{R}^n} m_{z(.)}(x) \tilde{z}(x) dx \qquad (7.27)$$

and

$$\frac{d}{d\theta} u_{z(.)+\theta\tilde{z}(.)}(x)|_{\theta=0} = \tilde{u}(x) \qquad (7.28)$$

with $\tilde{u}(x)$ solution of

$$A\tilde{u} + \alpha u = D\tilde{u}.g(x, \hat{v}(x, u_{z(.)}(x))) + \tilde{z}(x). \qquad (7.29)$$

Therefore,

$$\frac{d}{d\theta} \mathcal{K}(z(.) + \theta \tilde{z}(.))|_{\theta=0} = -\int_{\mathbb{R}^n} m_{z(.)}(x) \tilde{z}(x) dx + \alpha \int_{\mathbb{R}^n} m_0(x) \tilde{u}(x) dx.$$

Define next $\bar{m}_{z(.)}$ by

$$A^* \bar{m}_{z(.)} + \operatorname{div}\left(g(x, \hat{v}(x, u_{z(.)}(x))) \bar{m}_{z(.)}\right) + \alpha \bar{m}_{z(.)} = \alpha m_0 \qquad (7.30)$$

then from (7.29) to (7.30) we obtain easily

$$\alpha \int_{\mathbb{R}^n} m_0(x) \tilde{u}(x) dx = \int_{\mathbb{R}^n} \bar{m}_{z(.)}(x) \tilde{z}(x) dx$$

and thus we get

$$\frac{d}{d\theta} \mathcal{K}(z(.) + \theta \tilde{z}(.))|_{\theta=0} = \int_{\mathbb{R}^n} (\bar{m}_{z(.)}(x) - m_{z(.)}(x)) \tilde{z}(x) dx. \qquad (7.31)$$

If we express that the Frechet derivative is 0, then we must have $\bar{m}_{z(.)}(x) = m_{z(.)}(x)$. If we call $m(x)$ this common function, we have from (7.24)

$$z(x) = \frac{\partial \Phi(m)}{\partial m}(x).$$

If we call $u(x) = u_{z(.)}(x)$, then the pair u, m satisfies the system (7.1).

7.4   Approximate N Player Nash Equilibrium

We consider N independent Wiener processes, $w^i(t)$ and N random variables x_0^i, which are independent of the Wiener processes and identically distributed with density m_0. Consider local feedbacks (feedbacks on the individual states) $v^i(x^i)$. The evolution of the state of the player i is governed by

$$dx^i = g(x^i(t), v^i(x^i(t)))dt + \sigma(x^i(t))dw^i(t)$$

$$x^i(0) = x_0^i. \tag{7.32}$$

These trajectories are independent. The probability of the variable $x^i(t)$ is given by $p_{v^i(.)}(x^i, t)$ with $p_{v(.)}(x, t)$ the solution of

$$\frac{\partial p_{v(.)}}{\partial t} + A^* p_{v(.)} + \text{div}\,(g(x, v(x))p_{v(.)}) = 0$$

$$p_{v(.)}(x, 0) = m_0(x). \tag{7.33}$$

The objective functionals are defined by

$$\mathcal{J}^{N,i}(v(.)) = E \int_0^\infty \exp(-\alpha t)\left[f(x^i(t), v^i((x^i(t)))) \right.$$

$$\left. + f_0\left(x^i(t), \frac{\alpha}{N-1}\int_0^{+\infty}\exp(-\alpha s)\sum_{j=1\neq i}^N \delta_{x^j(s)}ds\right)\right] dt \tag{7.34}$$

Note the difference with respect to the time-dependent case; see (5.2).

This Nash game has an equilibrium composed of duplicates of a common feedback. Indeed, recall $\hat{v}(x, q)$ which attains the minimum of

$$f(x, v) + q.g(x, v)$$

and

$$H(x, q) = \inf[f(x, v) + q.g(x, v)]$$

$$H(x, m, q) = H(x, q) + f_0(x, m)$$

$$G(x, q) = g(x, \hat{v}(x, q)).$$

Next we define a system composed of a function $u_N(x)$ and stochastic processes $\hat{x}^1(.), \dots, \hat{x}^N$ which are solutions of

$$d\hat{x}^i = G(\hat{x}^i, Du_N(\hat{x}^i))dt + \sigma(\hat{x}^i(t))dw^i(t)$$

$$x^i(0) = x_0^i \tag{7.35}$$

$$Au_N + \alpha u_N = H(x, Du_N)$$

$$+ E f_0 \left(x, \frac{\alpha}{N-1} \int_0^{+\infty} \exp(-\alpha s) \sum_{j=2}^{N} \delta_{\hat{x}^j(s)} ds \right). \qquad (7.36)$$

If this system can be solved, we deduce feedbacks (dependent on N) $\hat{v}_N(x) = \hat{v}(x, Du_N(x))$. The copies $\hat{v}_N(x^i)$ form a Nash equilibrium.

Indeed, if we set $\hat{v}_N(.) = (\hat{v}_N(x^1), \ldots, \hat{v}_N(x^N))$, then

$$\mathcal{J}^{N,i}(\hat{v}_N(.)) = \int_{\mathbb{R}^n} u_N(x) m_0(x) dx. \qquad (7.37)$$

Let us next focus on player 1. Assume that player 1 uses a feedback control $v^1(x^1)$ and all other players use $\hat{v}_N(x^j)$, $j = 2, \ldots, N$. We use the notation

$$\mathcal{J}^{N,1}(v^1(.), \overline{\hat{v}_N^1}(.))$$

to denote the objective functional of player 1, when he uses the feedback $v^1(x^1)$ and the other players use the feedbacks $\hat{v}_N(x^j)$, $j = 2, \ldots, N$. The trajectory of player 1 is

$$dx^1 = g(x^1(t), v^1(x^1(t))) dt + \sigma(x^1(t)) dw^1(t)$$

$$x^1(0) = x_0^1$$

and the trajectories of other players remain $\hat{x}^j(t)$, $j = 2, \ldots, N$. By a standard verification argument, computing the Ito's differential of $u_N(x^1(t)) \exp(-\alpha t)$, one checks that

$$\int_{\mathbb{R}^n} u_N(x) m_0(x) dx \leq \mathcal{J}^{N,1}(v^1(.), \overline{\hat{v}_N^1}(.))$$

which proves the Nash equilibrium property.

It remains to compare (7.35) and (7.36) since $N \to +\infty$. Note that the probability densities of $\hat{x}^i(t)$ are identical and are defined by $\hat{p}_N(x,t)$ solution of

$$\frac{\partial \hat{p}_N}{\partial t} + A^* \hat{p}_N + \operatorname{div}(G(x, Du_N(x)) \hat{p}_N) = 0$$

$$\hat{p}_N(x,0) = m_0(x). \qquad (7.38)$$

We set

$$m_N(x) = \alpha \int_0^{+\infty} \exp(-\alpha t) \hat{p}_N(x,t) dt$$

then

$$A^* m_N + \operatorname{div}(G(x, Du_N(x)) m_N) + \alpha m_N = \alpha m_0. \qquad (7.39)$$

Consider now $\frac{\alpha}{N-1}\int_0^{+\infty}\exp(-\alpha s)\sum_{j=2}^N \delta_{\hat{x}^j(s)}ds$. It is a random measure on R^n. We can write for any continuous bounded function $\varphi(x)$ on $\mathbb{R}^n$

$$\left\langle \varphi, \frac{\alpha}{N-1}\int_0^{+\infty}\exp(-\alpha s)\sum_{j=2}^N \delta_{\hat{x}^j(s)}ds \right\rangle$$

$$= \frac{\alpha}{N-1}\int_0^{+\infty}\exp(-\alpha s)\sum_{j=2}^N \varphi(\hat{x}^j(s))ds$$

$$= \frac{\alpha}{N-1}\int_0^{+\infty}\exp(-\alpha s)\sum_{j=2}^N (\varphi(\hat{x}^j(s)) - E\varphi(\hat{x}^j(s)))ds$$

$$+ \int_{\mathbb{R}^n} \varphi(x)m_N(x)dx.$$

The random variables $\alpha\int_0^{+\infty}\exp(-\alpha s)(\varphi(\hat{x}^j(s)) - E\varphi(\hat{x}^j(s)))ds$ are independent, identically distributed with 0 mean. So from the law of large numbers, we obtain that

$$\frac{\alpha}{N-1}\int_0^{+\infty}\exp(-\alpha s)\sum_{j=2}^N (\varphi(\hat{x}^j(s)) - E\varphi(\hat{x}^j(s)))ds \to 0, \text{ a.s. as } N \to +\infty.$$

So, if we show that

$$\int_{\mathbb{R}^n} \varphi(x)m_N(x)dx \to \int_{\mathbb{R}^n} \varphi(x)m(x)dx$$

then we will get

$$\frac{\alpha}{N-1}\int_0^{+\infty}\exp(-\alpha s)\sum_{j=2}^N \delta_{\hat{x}^j(s)}ds \to m(x)dx, \text{ a.s.}$$

for the weak * topology of the set of measures on $\mathbb{R}^n$. If $f_0(x,m)$ is continuous in m for the weak * topology, then we will get

$$Ef_0\left(x, \frac{\alpha}{N-1}\int_0^{+\infty}\exp(-\alpha s)\sum_{j=2}^N \delta_{\hat{x}^j(s)}ds\right) \to f_0(x,m).$$

If we get estimates for u_N, m_N in Sobolev spaces, then we can extract a subsequence and the pair u_N, m_N will converge towards u, m, the solution of

$$Au + \alpha u = H(x, Du) + f_0(x,m)$$

$$A^*m + \text{div}\,(G(x,Du)m) + \alpha m = \alpha m_0. \tag{7.40}$$

This shows the approximation property.

Chapter 8
Different Populations

8.1 General Considerations

In preceding chapters, we considered a single population composed of a large number of individuals with identical behavior. In real situations, we will have several populations. The natural extension to the preceding developments is to obtain mean field equations for each population. A much more challenging situation will be to consider competing populations. This will be addressed in the next chapter. We discuss first the approach of multiclass agents, as described in [20, 22].

8.2 Multiclass Agents

We consider a more general situation than in [20, 22], which extends the model discussed in Sect. 5.4. Instead of functions $f(x,m,v), g(x,m,v), h(x,m), \sigma(x)$, we consider K functions $f_k(x,m,v), g_k(x,m,v), h_k(x,m), \sigma_k(x), k = 1,\ldots,K$. The index k represents some characteristics of the agents, and a class corresponds to one value of the characteristics. So there are K classes. In the model discussed previously, we considered a single class. In the sequel, when we consider an agent i, he or she will have a characteristics $\alpha^i \in (1,\ldots,K)$. Agents will be defined with upper indices, so $i = 1,\ldots,N$ with N very large. the value α^i is known information. The important assumption is

$$\frac{1}{N}\sum_{i=1}^{N}\mathbb{I}_{\alpha^i=k} \to \pi_k, \text{ as } N \to +\infty \tag{8.1}$$

and π_k is a probability distribution on the finite set of characteristics, which represents the probability that an agent has the characteristics k.

A. Bensoussan et al., *Mean Field Games and Mean Field Type Control Theory*,
SpringerBriefs in Mathematics, DOI 10.1007/978-1-4614-8508-7_8,
© Alain Bensoussan, Jens Frehse, Phillip Yam 2013

Generalizing the case of a single class, we define $a_k(x) = \frac{1}{2}\sigma_k(x)\sigma_k(x)^*$ and the operator

$$A_k\varphi(x) = -\text{tr}a_k(x)D^2\varphi(x).$$

We define Lagrangians—i.e., Hamiltonians indexed by k,—namely

$$L_k(x,m,v,q) = f_k(x,m,v) + q.g_k(x,m,v)$$

$$H_k(x,m,q) = \inf_v L_k(x,m,v,q)$$

and $\hat{v}_k(x,m,q)$ denote the minimizer in the definition of the Hamiltonian. We also define

$$G_k(x,m,q) = g_k(x,m,\hat{v}_k(x,m,q)).$$

Given a function $m(t)$ we consider the HJB equations indexed by k

$$-\frac{\partial u_k}{\partial t} + Au_k = H_k(x,m,Du_k)$$

$$u_k(x,T) = h_k(x,m(T)) \tag{8.2}$$

and the FP equations

$$\frac{\partial m_k}{\partial t} + A^*m_k + \text{div}\,(G_k(x,m,Du_k)m_k) = 0 \tag{8.3}$$

$$m_k(x,0) = m_{k0}(x) \tag{8.4}$$

in which the probability densities m_{k0} are given. A mean field game equilibrium for the multiclass agents problem is attained whenever

$$m(x,t) = \sum_{k=1}^{K} \pi_k m_k(x,t), \; \forall x,t. \tag{8.5}$$

To this mean field game we associate a Nash differential game for a large number of players, N. Considering independent Wiener processes $w^i(t)$, $i = 1,\dots,N$, we define the state equations

$$dx^i(t) = g_{\alpha^i}\left(x^i(t), \frac{1}{N-1}\sum_{j=1\neq i}\delta_{x^j(t)}, v^i(x^i(t))\right)dt + \sigma_{\alpha^i}(x^i(t))dw^i(t)$$

$$x^i(0) = x_0^i, \tag{8.6}$$

where the random variables x_0^i are independent, with probability density $m_{\alpha^i 0}(x)$. In the state equation $v^i(x^i(t))$ denotes a feedback on the agent's state. We define the cost functional associated with player i

$$\mathcal{J}^{N,i}(v(.)) = E\left[\int_0^T f_{\alpha^i}\left(x^i(t), \frac{1}{N-1}\sum_{j=1\neq i}^N \delta_{x^j(t)}, v^i(x^i(t))\right) dt\right.$$

$$\left. + h_{\alpha^i}\left(x^i(T), \frac{1}{N-1}\sum_{j=1\neq i}^N \delta_{x^j(T)}\right)\right]. \tag{8.7}$$

The objective is to obtain an approximate Nash equilibrium. We consider the feedbacks

$$\hat{v}_k(x,t) = \hat{v}_k(x, m(t), Du_k(x,t))$$

and player i will use the local feedback

$$\hat{v}^i(x^i,t) = \hat{v}_{\alpha^i}(x^i,t). \tag{8.8}$$

We want to show that these feedbacks constitute an approximate Nash equilibrium for the differential game with N players; see (8.6) and (8.7). The admissible feedbacks will be Lipschitz feedbacks. We first consider the trajectories

$$d\hat{x}^i(t) = g_{\alpha^i}(\hat{x}^i(t), m(t), \hat{v}^i(\hat{x}^i(t),t))dt + \sigma_{\alpha^i}(\hat{x}^i(t))dw^i(t)$$

$$\hat{x}^i(0) = x_0^i. \tag{8.9}$$

In this equation, only player i is involved. We then consider the cost

$$J_{\alpha^i}(\hat{v}^i(.)) = E\left[\int_0^T f_{\alpha^i}(\hat{x}^i(t), m(t), \hat{v}^i(\hat{x}^i(t),t))dt\right.$$

$$\left. + h_{\alpha^i}(\hat{x}^i(T), m(T))\right]. \tag{8.10}$$

Again, this cost functional involves only player i. By standard results of dynamic programming, we can check that

$$J_{\alpha^i}(\hat{v}^i(.)) = \int u_{\alpha^i}(x,0)m(x,0)dx. \tag{8.11}$$

Consider next the trajectories of the N players when they simultaneously apply the feedbacks $\hat{v}^i(x^i,t)$. They are defined as follows

$$d\hat{y}^i(t) = g_{\alpha^i}\left(\hat{y}^i(t), \frac{1}{N-1}\sum_{j=1\neq i}^{N}\delta_{\hat{y}^j(t)}, \hat{v}^i(\hat{y}^i(t),t)\right)dt + \sigma_{\alpha^i}(\hat{y}^i(t))dw^i(t)$$

$$\hat{y}^i(0) = x_0^i \tag{8.12}$$

and the corresponding cost functional is given by

$$\mathcal{J}^{N,i}(\hat{v}(.))) = E\left[\int_0^T f_{\alpha^i}\left(\hat{y}^i(t), \frac{1}{N-1}\sum_{j=1\neq i}^{N}\delta_{\hat{y}^j(t)}, \hat{v}^i(\hat{y}^i(t),t)\right)dt \right.$$

$$\left. + h_{\alpha^i}\left(\hat{y}^i(T), \frac{1}{N-1}\sum_{j=1\neq i}^{N}\delta_{\hat{y}^j(T)}\right)\right], \tag{8.13}$$

where $\hat{v}(.) = (\hat{v}^1(.), \ldots, \hat{v}^N(.))$. We claim that

$$\mathcal{J}^{N,i}(\hat{v}(.))) = J_{\alpha^i}(\hat{v}^i(.)) + O\left(\frac{1}{\sqrt{N}}\right). \tag{8.14}$$

We proceed as in Sect. 5.4. We show that $\hat{x}^i(t)$ is an approximation of $\hat{y}^i(t)$. We write

$$d(\hat{y}^i(t) - \hat{x}^i(t))$$

$$= \left[g_{\alpha^i}\left(\hat{y}^i(t), \frac{1}{N-1}\sum_{j=1\neq i}^{N}\delta_{\hat{y}^j(t)}, \hat{v}^i(\hat{y}^i(t),t)\right) - g_{\alpha^i}\left(\hat{x}^i(t), \frac{1}{N-1}\sum_{j=1\neq i}^{N}\delta_{\hat{y}^j(t)}, \hat{v}^i(\hat{x}^i(t),t)\right)\right.$$

$$+ g_{\alpha^i}\left(\hat{x}^i(t), \frac{1}{N-1}\sum_{j=1\neq i}^{N}\delta_{\hat{y}^j(t)}, \hat{v}^i(\hat{x}^i(t),t)\right) - g_{\alpha^i}\left(\hat{x}^i(t), \frac{1}{N-1}\sum_{j=1\neq i}^{N}\delta_{\hat{x}^j(t)}, \hat{v}^i(\hat{x}^i(t),t)\right)$$

$$\left. + g_{\alpha^i}\left(\hat{x}^i(t), \frac{1}{N-1}\sum_{j=1\neq i}^{N}\delta_{\hat{x}^j(t)}, \hat{v}^i(\hat{x}^i(t),t)\right) - g_{\alpha^i}(\hat{x}^i(t), m(t), \hat{v}^i(\hat{x}^i(t),t))\right]dt$$

$$+ (\sigma_{\alpha^i}(\hat{y}^i(t)) - \sigma_{\alpha^i}(\hat{x}^i(t)))dw^i$$

$$\hat{y}^i(0) - \hat{x}^i(0) = 0.$$

With appropriate Lipschitz assumptions, we control all terms with the difference $\sum_{i=1}^{N}|\hat{y}^i(t) - \hat{x}^i(t)|^2$, except the term

$$g_{\alpha^i}\left(\hat{x}^i(t), \frac{1}{N-1}\sum_{j=1\neq i}^{N}\delta_{\hat{x}^j(t)}, \hat{v}^i(\hat{x}^i(t),t)\right) - g_{\alpha^i}(\hat{x}^i(t), m(t), \hat{v}^i(\hat{x}^i(t),t))$$

which is the driving term converging to 0 as $N \to +\infty$. Indeed, we have to show that the random measure on $\mathbb{R}^n$ defined by $\frac{1}{N-1}\sum_{j=1\neq i}^{N}\delta_{\hat{x}^j(t)}$ converges for any t towards the deterministic measure $m(x,t)dx$, a.s.

But if φ is continuous and bounded on $\mathbb{R}^n$ we can write

$$\left\langle \varphi, \frac{1}{N-1} \sum_{j=1 \neq i}^{N} \delta_{\hat{x}^j(t)} \right\rangle = \frac{1}{N-1} \sum_{j=1 \neq i}^{N} \varphi(\hat{x}^j(t))$$

$$= \frac{1}{N-1} \sum_{j=1 \neq i}^{N} (\varphi(\hat{x}^j(t)) - E\varphi(\hat{x}^j(t))) + \frac{1}{N-1} \sum_{j=1 \neq i}^{N} E\varphi(\hat{x}^j(t)).$$

But the random variables $\varphi(\hat{x}^j(t)) - E\varphi(\hat{x}^j(t))$ are independent and identically distributed with mean 0. From the law of large numbers it follows that

$$\frac{1}{N-1} \sum_{j=1 \neq i}^{N} (\varphi(\hat{x}^j(t)) - E\varphi(\hat{x}^j(t))) \to 0, \text{ a.s. for fixed } i, \text{ as } N \to \infty.$$

Next the probability density of $\hat{x}^j(t)$ is $m_{\alpha^j}(t)$. Therefore,

$$\frac{1}{N-1} \sum_{j=1 \neq i}^{N} E\varphi(\hat{x}^j(t)) = \frac{1}{N-1} \sum_{j=1 \neq i}^{N} \int_{\mathbb{R}^n} \varphi(x) m_{\alpha^j}(x,t) dx$$

$$= \sum_{k=1}^{K} \int \varphi(x) m_k(x,t) \frac{1}{N-1} \sum_{j=1 \neq i}^{N} \mathbb{I}_{\alpha^j = k} dx$$

and from the assumption (8.1) we obtain that

$$\frac{1}{N-1} \sum_{j=1 \neq i}^{N} E\varphi(\hat{x}^j(t)) \to \sum_{k=1}^{K} \int \varphi(x) m_k(x,t) \pi_k dx = \int \varphi(x) m(x,t) dx$$

which proves the convergence of $\frac{1}{N-1} \sum_{j=1 \neq i}^{N} \delta_{\hat{x}^j(t)}$ to $m(x,t)dx$ a.s. The result follows from an assumption of continuity of $g_k(x,m,v)$ in m, with respect to the weak * convergence of probability measures.

With this in hand we can compare $\mathcal{J}^{N,i}(\hat{v}(.))$ with $J_{\alpha^i}(\hat{v}^i(.))$ as done for (5.36) in Sect. 5.4. This proves (8.14).

We next focus on player 1. Suppose this player uses a different local feedback $v^1(x^1)$ that is Lipschitz. The other players use $\hat{v}^i(x^i), i = 2,\ldots,N$. We call this set of controls $\tilde{v}(.) = (v^1(.), \hat{v}^2(.), \ldots, \hat{v}^N(.))$ and we use it in the differential game; see (8.6) and (8.7). We note the corresponding states by $(y^1(.),\ldots,y^N(.))$. We get the equations

$$dy^1(t) = g_{\alpha^1} \left(y^1(t), \frac{1}{N-1} \sum_{j=1 \neq i}^{N} \delta_{y^j(t)}, v^1(y^1(t)) \right) dt + \sigma_{\alpha^1}(y^1(t)) dw^1(t)$$

$$y^1(0) = x_0^1 \tag{8.15}$$

$$dy^i(t) = g_{\alpha^i}\left(y^i(t), \frac{1}{N-1}\sum_{j=1\neq i}\delta_{y^j(t)}, \hat{v}^i(y^i(t))\right)dt + \sigma_{\alpha^i}(y^i(t))dw^i(t)$$

$$y^i(0) = x_0^i \qquad\qquad\qquad\qquad\qquad\qquad\qquad\qquad\qquad (8.16)$$

for $i = 2, \ldots, N$.

We also consider the state evolution of the first player, when he uses the feedback $v^1(.)$ and the measure is $m(t)$, given by (8.5), namely

$$dx^1(t) = g_{\alpha^1}(x^1(t), m(t), v^1(x^1(t)))dt + \sigma_{\alpha^1}(x^1(t))dw^1(t)$$

$$x^1(0) = x_0^1 \qquad\qquad\qquad\qquad\qquad\qquad\qquad\qquad\qquad (8.17)$$

then we show that $x^1(t)$ approximates $y^1(t)$ and $\hat{x}^i(t)$ approximates $y^i(t)$, for $i \geq 2$. The proof is very similar to that of Sect. 5.4 and what has been done above. We arrive at

$$\mathcal{J}^{N,1}(\tilde{v}(.)) = J_{\alpha^1}(v^1(.)) + O\left(\frac{1}{\sqrt{N}}\right)$$

with

$$J_{\alpha^1}(v^1(.)) = E\left[\int_0^T f_{\alpha^1}(x^1(t), m(t), v^1(x^1(t), t))dt\right.$$

$$\left. + h_{\alpha^1}(x^1(T), m(T))\right]. \qquad\qquad\qquad (8.18)$$

By standard control theory we have $J_{\alpha^1}(v^1(.)) \geq J_{\alpha^1}(\hat{v}^1(.))$ and thus again

$$\mathcal{J}^{N,1}(\tilde{v}(.)) \geq J_{\alpha^1}(\hat{v}^1(.)) - O\left(\frac{1}{\sqrt{N}}\right)$$

and we obtain the approximate Nash equilibrium for a large number of multiclass agents.

8.3 Major Player

8.3.1 General Theory

We consider here a problem initiated by Huang [19]. In this paper only the LQ case is considered. In a recent paper, Nourian and Caines [31] have studied a nonlinear mean field game with a major player. In both papers, there is a simplification in the coupling between the major player and the representative agent. We will describe here the problem in full generality and explain the simplification in [31].

The new element is that, besides the representative agent, there is a major player. This major player influences directly the mean field term. Since the mean field term also impacts the major player, he or she will takes this into account to define any decisions made. On the other hand, the mean field term can no longer be deterministic, since it depends on the major player's decisions. This coupling creates new difficulties.

We introduce the following state evolution for the major player

$$dx_0 = g_0(x_0(t), m(t), v_0(t))dt + \sigma_0(x_0)dw_0$$
$$x_0(0) = \xi_0. \tag{8.19}$$

We assume that $x_0(t) \in \mathbb{R}^{n_0}$, $v_0(t) \in \mathbb{R}^{d_0}$. The process $w_0(t)$ is a standard Wiener process with values in $\mathbb{R}^{k_0}$ and ξ_0 is a random variable in $\mathbb{R}^{n_0}$ independent of the Wiener process. The process $m(t)$ is the mean field term, with values in the space of probabilities on $\mathbb{R}^n$. This term will come from the decisions of the representative agent. However, it will be linked to $x_0(t)$ since the major player influences the decision of the representative agent. If we define the filtration

$$\mathcal{F}^{0t} = \sigma(\xi_0, w_0(s), s \leq t) \tag{8.20}$$

then $m(t)$ is a process adapted to $\mathcal{F}^{0t}$. But it is not external, since it is assumed in the above works. We will describe the link with the state x_0 in analyzing the representative agent problem. The control $v_0(t)$ is also adapted to $\mathcal{F}^{0t}$. The objective functional of the major player is

$$J_0(v_0(.)) = E\left[\int_0^T f_0(x_0(t), m(t), v_0(t))dt + h_0(x_0(T), m(T))\right]. \tag{8.21}$$

The functions g_0, f_0, σ_0, h_0 are deterministic. We do not specify the assumptions, since our treatment is formal.

We turn now to the representative agent problem. The state $x(t) \in \mathbb{R}^n$ and the control $v(t) \in \mathbb{R}^d$. We have the evolution

$$dx = g(x(t), x_0(t), m(t), v(t))dt + \sigma(x(t))dw$$
$$x(0) = \xi \tag{8.22}$$

in which $w(t)$ is a standard Wiener process with values in $\mathbb{R}^k$ and ξ is a random variable with values in $\mathbb{R}^n$ independent of $w(.)$. Moreover, $\xi, w(.)$ are independent of $\xi_0, w_0(.)$. We define

$$\mathcal{F}^t = \sigma(\xi, w(s), s \leq t) \tag{8.23}$$
$$\mathcal{G}^t = \mathcal{F}^{0t} \cup \mathcal{F}^t. \tag{8.24}$$

The control $v(t)$ is adapted to $\mathcal{G}^t$. The objective functional of the representative agent is defined by

$$J(v(.),x_0(.),m(.)) = E\left[\int_0^T f(x(t),x_0(t),m(t),v(t))dt + h(x(T),x_0(T),m(T)))\right].$$
(8.25)

Conversely to the major player problem, in the representative agent problem the processes $x_0(.),m(.)$ are external. This explains the difference of notation between (8.21) and (8.25). In (8.21), $m(t)$ depends on $x_0(.)$.

The representative agent's problem is similar to the standard situation of Sect. 2.1 except for the presence of $x_0(t)$.

We begin by limiting the class of controls for the representative agent to belong to feedbacks $v(x,t)$ random fields adapted to $\mathcal{F}^{0t}$. The corresponding state, solution of (8.22) is denoted by $x_{v(.)}(t)$. Of course, this process depends also of $x_0(t),m(t)$. Note that $x_0(t),m(t)$ is independent from $\mathcal{F}^t$, therefore the conditional probability density of $x_{v(.)}(t)$ given the filtration $\cup_t \mathcal{F}^{0t}$ is the solution of the FP equation with random coefficients

$$\frac{\partial p_{v(.)}}{\partial t} + A^* p_{v(.)} + \operatorname{div}(g(x,x_0(t),m(t),v(x,t))p_{v(.)}) = 0$$

$$p_{v(.)}(x,0) = \varpi(x)$$
(8.26)

in which $\varpi(x)$ is the density probability of ξ. We can then rewrite the objective functional $J(v(.),x_0(.),m(.))$ as follows

$$J(v(.),x_0(.),m(.)) = E\left[\int_0^T \int_{\mathbb{R}^n} p_{v(.),x_0(.),m(.)}(x,t)f(x,x_0(t),m(t),v(x,t))dxdt\right.$$

$$\left. + \int_{\mathbb{R}^n} p_{v(.),x_0(.),m(.)}(x,T)h(x,x_0(T),m(T))dx\right].$$
(8.27)

We can give an expression for this functional. Introduce the random field $\chi_{v(.)}(x,t)$ solution of the stochastic backward PDE:

$$-\frac{\partial \chi_{v(.)}}{\partial t} + A\chi_{v(.)} = f(x,x_0(t),m(t),v(x,t)) + g(x,x_0(t),m(t),v(x,t)).D\chi_{v(.)}$$

$$\chi_{v(.)}(x,T) = h(x,x_0(T),m(T))$$
(8.28)

then we can assert that

$$\int_0^T \int_{\mathbb{R}^n} p_{v(.),x_0(.),m(.)}(x,t)f(x,x_0(t),m(t),v(x,t))dxdt$$

$$+ \int_{\mathbb{R}^n} p_{v(.),x_0(.),m(.)}(x,T)h(x,x_0(T),m(T))dx = \int_{\mathbb{R}^n} \chi_{v(.)}(x,0)\varpi(x)dx$$

so

$$J(v(.),x_0(.),m(.)) = \int_{\mathbb{R}^n} \varpi(x) E \chi_{v(.)}(x,0) dx. \qquad (8.29)$$

Now define

$$u_{v(.)}(x,t) = E^{\mathcal{F}^{0t}} \chi_{v(.)}(x,t).$$

From (8.28) we can assert that

$$-E^{\mathcal{F}^{0t}} \frac{\partial \chi_{v(.)}}{\partial t} + A u_{v(.)} = f(x,x_0(t),m(t),v(x,t)) + g(x,x_0(t),m(t),v(x,t)).Du_{v(.)}$$

$$u_{v(.)}(x,T) = h(x,x_0(T),m(T)). \qquad (8.30)$$

On the other hand

$$u_{v(.)}(x,t) - \int_0^t E^{\mathcal{F}^{0s}} \frac{\partial \chi_{v(.)}}{\partial s}(x,s) ds$$

is a $\mathcal{F}^{0t}$ martingale. Therefore, we can write

$$u_{v(.)}(x,t) - \int_0^t E^{\mathcal{F}^{0s}} \frac{\partial \chi_{v(.)}}{\partial s}(x,s) ds = u_{v(.)}(x,0) + \int_0^t K_{v(.)}(x,s) dw_0(s),$$

where $K_{v(.)}(x,s)$ is $\mathcal{F}^{0s}$ measurable, and uniquely defined. It is then easy to check that the random field $u_{v(.)}(x,t)$ is a solution of the backward stochastic PDE (SPDE):

$$-\partial_t u_{v(.)}(x,t) + A u_{v(.)}(x,t) dt = f(x,x_0(t),m(t),v(x,t)) dt$$

$$+ g(x,x_0(t),m(t),v(x,t)).Du_{v(.)}(x,t) dt$$

$$- K_{v(.)}(x,t) dw_0(t)$$

$$u_{v(.)}(x,T) = h(x,x_0(T),m(T)). \qquad (8.31)$$

From (8.29) we get immediately

$$J(v(.),x_0(.),m(.)) = \int_{\mathbb{R}^n} \varpi(x) E u_{v(.)}(x,0) dx. \qquad (8.32)$$

To express a necessary condition of optimality, we have to compute the Gâteaux differential

$$\frac{d}{d\theta} J(v(.) + \theta \tilde{v}(.),x_0(.),m(.)).$$

We can state

$$\frac{d}{d\theta} J(v(.) + \theta \tilde{v}(.),x_0(.),m(.)) = \int_{\mathbb{R}^n} \varpi(x) E \tilde{u}(x,0) dx \qquad (8.33)$$

with

$$-\partial_t \tilde{u}(x,t) + A\tilde{u}(x,t)dt$$
$$= g(x,x_0(t),m(t),v(x,t)).D\tilde{u}(x,t)dt$$
$$+ \frac{\partial L}{\partial v}(x,x_0(t),m(t),v(x,t),Du_{v(.)}(x,t))\tilde{v}(x,t)dt - \tilde{K}(x,t)dw_0(t)$$
$$\tilde{u}(x,T) = 0, \qquad (8.34)$$

where

$$L(x,x_0,m,v,q) = f(x,x_0,m,v) + q.g(x,x_0,m,v). \qquad (8.35)$$

Combining (8.26) and (8.34), we obtain

$$\frac{d}{d\theta}J(v(.) + \theta\tilde{v}(.),x_0(.),m(.))$$
$$= E\int_0^T \int_{\mathbb{R}^n} p_{v(.)}(x,t)\frac{\partial L}{\partial v}(x,x_0(t),m(t),v(x,t),Du_{v(.)}(x,t))\tilde{v}(x,t)dxdt. \qquad (8.36)$$

So the optimal feedback $\hat{v}(x,t)$ must satisfy

$$\frac{\partial L}{\partial v}(x,x_0(t),m(t),\hat{v}(x,t),Du_{\hat{v}(.)}(x,t)) = 0, \text{ a.e.} x,t, \text{ a.s.} \qquad (8.37)$$

As usual we consider that the function L achieves a minimum in v, denoted by $\hat{v}(x,x_0,m,q)$ and define

$$H(x,x_0,m,q) = L(x,x_0,m,\hat{v}(x,x_0,m,q),q). \qquad (8.38)$$

Setting $u(x,t) = u_{\hat{v}(.)}(x,t)$, $K(x,t) = K_{\hat{v}(.)}(x,t)$ we obtain the stochastic HJB equation

$$-\partial_t u(x,t) + Au(x,t)dt = H(x,x_0(t),m(t),Du)dt - K(x,t)dw_0$$
$$u(x,T) = h(x,x_0(T),m(T)) \qquad (8.39)$$

and

$$\hat{v}(x,t) = \hat{v}(x,x_0(t),m(t),Du(x,t)). \qquad (8.40)$$

We next have to express the mean field game condition

$$m(t) = p_{\hat{v}(.),x_0(.),m(.)}(.,t).$$

Setting

$$G(x,x_0,m,q) = g(x,x_0,m,\hat{v}(x,x_0,m,q)) \tag{8.41}$$

we obtain from (8.26) the FP equation

$$\frac{\partial m}{\partial t} + A^*m + \mathrm{div}(G(x,x_0(t),m(t),Du(x,t))m) = 0$$

$$m(x,0) = \varpi(x). \tag{8.42}$$

The coupled pair of HJB-FP equations, (8.39) and (8.42), allow is to define the reaction function of the representative agent to the trajectory $x_0(.)$ of the major player. One defines the random fields $u(x,t)$ and $m(x,t)$, and the optimal feedback is given by (8.40).

Consider now the problem of the major player. In [31] and also [19] for the LQ case it is limited to (8.19) and (8.21) since $m(t)$ is external. However, since $m(t)$ is coupled to $x_0(t)$ through (8.39) and (8.42), one cannot consider $m(t)$ as external, unless limiting the decision of the major player. So, in fact, the major player has to consider three state equations: (8.19), (8.39), and (8.42). For a given $v_0(.)$ adapted to $\mathcal{F}^{0t}$ we associate $x_{0,v_0(.)}(.), u_{v_0(.)}(.,.), m_{v_0(.)}(.,.)$ as the solution of the system (8.19), (8.39), and (8.42). We may drop the index $v_0(.)$ when the context is clear. We need to compute the Gâteaux differentials

$$\tilde{x}_0(t) = \frac{d}{d\theta}x_{0,v_0(.)+\theta v_0(.)}(t)|_{\theta=0}, \quad \tilde{u}(x,t) = \frac{d}{d\theta}u_{v_0(.)+\theta\tilde{v}_0(.)}(x,t)|_{\theta=0}$$

$$\tilde{m}(x,t) = \frac{d}{d\theta}m_{v_0(.)+\theta\tilde{v}_0(.)}(x,t)|_{\theta=0}.$$

They are solutions of the following equations

$$d\tilde{x}_0 = \left[g_{0,x_0}(x_0(t),m(t),v_0(t))\tilde{x}_0(t)\right.$$

$$+ \int \frac{\partial g_0}{\partial m}(x_0(t),m(t),v_0(t))(\xi)\tilde{m}(\xi,t)d\xi$$

$$\left. + g_{0,v_0}(x_0(t),m(t),v_0(t))\tilde{v}_0(t)\right]dt$$

$$+ \sum_{l=1}^{k_0}\sigma_{0l,x_0}(x_0(t))\tilde{x}_0(t)dw_{0l} \tag{8.43}$$

$$\tilde{x}_0(0) = 0$$

$$-\partial_t \tilde{u}(x,t) + A\tilde{u}(x,t)dt$$

$$= \left[H_{x_0}(x,x_0(t),m(t),Du(x,t))\tilde{x}_0(t) + G(x,x_0(t),m(t),Du(x,t)).D\tilde{u}(x,t) \right.$$

$$\left. + \int \frac{\partial H}{\partial m}(x,x_0(t),m(t),Du(x,t))(\xi)\tilde{m}(\xi,t)d\xi \right] dt - \tilde{K}(x,t)dw_0(t)$$

$$\tilde{u}(x,T) = h_{x_0}(x,x_0(T),m(T))\tilde{x}_0(T) + \int \frac{\partial h}{\partial m}(x,x_0(T),m(T))(\xi)\tilde{m}(\xi,T)d\xi$$

$$(8.44)$$

$$\frac{\partial}{\partial t}\tilde{m}(x,t) + A^*\tilde{m}(x,t) + \mathrm{div}(G(x,x_0(t),m(t),Du(x,t))\tilde{m})$$

$$+\mathrm{div}\left(\left(G_{x_0}(x,x_0(t),m(t),Du(x,t))\tilde{x}_0(t)\right.\right.$$

$$+ \int \frac{\partial G}{\partial m}(x,x_0(t),m(t),Du(x,t))(\xi)\tilde{m}(\xi,t)d\xi$$

$$\left.\left. + G(x,x_0(t),m(t),Du(x,t)).D\tilde{u}(x,t)\right)m(x,t)\right) = 0 \qquad (8.45)$$

$$\tilde{m}(x,0) = 0.$$

We then can assert that

$$\frac{d}{d\theta}J_0(v_0(.) + \theta\tilde{v}_0(.))|_{\theta=0}$$

$$= E\int_0^T \left[f_{0,x_0}(x_0(t),m(t),v_0(t))\tilde{x}_0(t) + \int \frac{\partial f_0}{\partial m}(x_0(t),m(t),v_0(t))(\xi)\tilde{m}(\xi,t)d\xi \right.$$

$$\left. + f_{0,v_0}(x_0(t),m(t),v_0(t))\tilde{v}_0(t) \right] dt + E\left[h_{0,x_0}(x_0(T),m(T))\tilde{x}_0(T) \right.$$

$$\left. + \int \frac{\partial h_0}{\partial m}(x_0(T),m(T))(\xi)\tilde{m}(\xi,T)d\xi \right]. \qquad (8.46)$$

We then introduce the process $p(t)$ and random fields $\eta(x,t)$, $\zeta(x,t)$ as solutions of the SDE and SPDE:

$$-dp = \left[g^*_{0,x_0}(x_0(t),m(t),v_0(t))p(t) + f_{0,x_0}(x_0(t),m(t),v_0(t)) \right.$$

$$\left. + \int G^*_{x_0}(x,x_0(t),m(t),Du(x,t))D\eta(x,t)m(x,t)dx \right.$$

$$+ \int \zeta(x,t) H_{x_0}(x,x_0(t),m(t),Du(x,t)dx \Bigg] dt$$

$$- \sum_{l=1}^{k_0} q_l dw_{0l} + \sum_{l=1}^{k_0} \sigma^*_{0l,x_0}(x_0(t)) q_l dt. \qquad (8.47)$$

$$p(T) = h_{0,x_0}(x_0(T),m(T)) + \int \zeta(x,T) h_{x_0}(x,x_0(T),m(T)) dx$$

$$- \partial_t \eta + A\eta(x,t) dt = \Bigg[\frac{\partial g_0}{\partial m}(x_0(t),m(t),v_0(t))(x) . p(t)$$

$$+ D\eta(x,t) . G(x,x_0(t),m(t),Du(x,t))$$

$$+ \int D\eta(\xi,t) . \frac{\partial G}{\partial m}(\xi,x_0(t),m(t),Du(\xi,t))(x) m(\xi,t) d\xi$$

$$+ \int \zeta(\xi,t) \frac{\partial H}{\partial m}(\xi,x_0(t),m(t),Du(\xi,t))(x) d\xi$$

$$+ \frac{\partial f_0}{\partial m}(x_0(t),m(t),v_0(t))(x) \Bigg] dt - \sum_l \mu_l(x,t) dw_{0l}(t) \qquad (8.48)$$

$$\eta(x,T) = \frac{\partial h_0}{\partial m}(x_0(T),m(T))(x) + \int \zeta(\xi,T) \frac{\partial h}{\partial m}(\xi,x_0(T),m(T))(x) d\xi$$

$$\frac{\partial \zeta}{\partial t} + A^* \zeta(x,t) + \mathrm{div}\,(G(x,x_0(t),m(t),Du(x,t)) \zeta(x,t))$$

$$+ \mathrm{div}(G^*_q(x,x_0(t),m(t),Du(x,t)) D\eta(x,t) m(x,t)) = 0 \qquad (8.49)$$

$$\zeta(x,0) = 0.$$

Thanks to these processes we can write (8.46) as follows

$$\frac{d}{d\theta} J_0(v_0(.) + \theta \tilde{v}_0(.))|_{\theta=0} = E \int_0^T \big(f_{0,v_0}(x_0(t),m(t),v_0(t))$$

$$+ p(t)^* g_{0,v_0}(x_0(t),m(t),v_0(t)) \big) \tilde{v}_0(t) dt \qquad (8.50)$$

and writing the necessary condition that this Gâteaux differential must be equal to 0, we obtain

$$v_0(t) \text{ minimizes } f_0(x_0(t),m(t),v_0) + p(t) \cdot g_0(x_0(t),m(t),v_0) \text{ in } v_0.$$

We introduce the notation

$$H_0(x_0,m,p) = \inf_{v_0}[f_0(x_0,m,v_0) + p \cdot g_0(x_0,m,v_0)]$$

$$\hat{v}_0(x_0,m,p) \quad \text{minimizes the expression in brackets}$$

$$G_0(x,m,p) = g_0(x_0,m,\hat{v}_0(x_0,m,p))$$

so we can write from (8.47)–(8.49)

$$-dp = \left[H_{0,x_0}(x_0(t),m(t),p(t)) + \sum_{l=1}^{k_0} \sigma^*_{0l,x_0}(x_0(t))q_l(t) \right.$$

$$+ \int G^*_{x_0}(x,x_0(t),m(t),Du(x,t))D\eta(x,t)m(x,t)dx$$

$$\left. + \int \zeta(x,t)H_{x_0}(x,x_0(t),m(t),Du(x,t)dx \right] dt - \sum_{l=1}^{k_0} q_l dw_{0l} \quad (8.51)$$

$$p(T) = h_{0,x_0}(x_0(T),m(T)) + \int \zeta(x,T)h_{x_0}(x,x_0(T),m(T))dx$$

$$-\partial_t \eta + A\eta(x,t)dt = \left[\frac{\partial H_0}{\partial m}(x_0(t),m(t),p(t))(x) \right.$$

$$+ D\eta(x,t) \cdot G(x,x_0(t),m(t),Du(x,t))$$

$$+ \int D\eta(\xi,t) \cdot \frac{\partial G}{\partial m}(\xi,x_0(t),m(t),Du(\xi,t))(x)m(\xi,t)d\xi$$

$$\left. + \int \zeta(\xi,t) \frac{\partial H}{\partial m}(\xi,x_0(t),m(t),Du(\xi,t))(x)d\xi \right] dt - \sum_l \mu_l(x,t)dw_{0l}(t)$$

$$(8.52)$$

$$\eta(x,T) = \frac{\partial h_0}{\partial m}(x_0(T),m(T))(x) + \int \zeta(\xi,T)\frac{\partial h}{\partial m}(\xi,x_0(T),m(T))(x)d\xi$$

$$\frac{\partial \zeta}{\partial t} + A^* \zeta(x,t) + \text{div} \left(G(x,x_0(t),m(t),Du(x,t))\zeta(x,t) \right)$$

$$+ \text{div}(G_q^*(x,x_0(t),m(t),Du(x,t))D\eta(x,t)m(x,t)) = 0 \qquad (8.53)$$

$$\zeta(x,0) = 0.$$

Next $x_0(t)$ satisfies

$$dx_0 = G_0(x_0(t),m(t),p(t))dt + \sigma_0(x_0(t))dw_0$$

$$x_0(0) = \xi_0. \qquad (8.54)$$

So, in fact, the complete solution is provided by the six equations—(8.54), (8.51), (8.39), (8.53), (8.42), and (8.52)—and the feedback of the representative agent and the control of the major player are given by (8.40) and

$$\hat{v}_0(t) = \hat{v}_0(x_0(t),m(t),p(t)). \qquad (8.55)$$

If we follow the approach of Nourian–Caines [31], then the major player considers $m(t)$ as external. In that case,

$$\eta(x,t), \ \zeta(x,t) = 0$$

and thus the six equations reduce to four, namely (8.54), (8.39), (8.42), and

$$-dp = \left(H_{0,x_0}(x_0(t),m(t),p(t)) + \sum_{l=1}^{k_0} \sigma_{0l,x_0}^*(x_0(t))q_l(t) \right) dt - \sum_{l=1}^{k_0} q_l dw_{0l} \qquad (8.56)$$

$$p(T) = h_{0,x_0}(x_0(T),m(T)).$$

In fact, in this case the control problem of the major player is simply (8.19) and (8.21), in which $m(t)$ is a given process adapted to $\mathcal{F}^{0t}$. This is a standard stochastic control problem, except that the drift is random and adapted. We can apply the theory of Peng [32] and introduce the HJB equation

$$-\partial_t u_0(x_0,t) + A_0 u_0(x_0,t)dt = \left(H_0(x_0,m(t),Du_0(x_0,t)) \right.$$

$$+ \sum_{l=1}^{k_0} \sum_{i=1}^{n_0} \sigma_{0il}(x_0)K_{0l,i}(x_0,t) \Bigg) dt$$

$$- \sum_{l=1}^{k_0} K_{0l}(x_0,t)dw_0(t) \qquad (8.57)$$

$$u_0(x_0,T) = h_0(x_0,m(T))$$

in which

$$a_0(x_0) = \frac{1}{2}\sigma_0(x_0)\sigma_0^*(x_0)$$

$$A_0\varphi(x_0) = -\operatorname{tr} a_0(x_0)D^2\varphi(x_0). \tag{8.58}$$

The solution of the major player problem is then defined by the four equations (8.54), (8.39), (8.42), and (8.57). Naturally we have the relation

$$p_i(t) = u_{0,i}(x_0(t),t), \quad q_{il}(t) = K_{0l,i}(x_0(t),t) + \sum_{j=1}^{n_0} u_{0,ij}(x_0(t),t)\sigma_{0jl}(x_0(t)) \tag{8.59}$$

in which

$$u_{0,i}(x_0,t) = \frac{\partial u_0}{\partial x_{0i}}(x_0,t), \quad K_{0l,i}(x_0,t) = \frac{\partial K_{0l}}{\partial x_{0i}}(x_0,t), \quad u_{0,ij}(x_0,t) = \frac{\partial^2 u_0}{\partial x_{0i}\partial x_{0j}}(x_0,t).$$

8.3.2 Linear Quadratic Case

We now consider the linear quadratic case and apply the general theory above. We assume

$$g_0(x_0,m,v_0) = A_0x_0 + B_0v_0 + F_0\int \xi m(\xi)d\xi$$

$$\sigma_0(x_0) = \sigma_0 \tag{8.60}$$

$$f_0(x_0,m,v_0) = \frac{1}{2}\left[\left(x_0 - H_0\int \xi m(\xi)d\xi - \gamma_0\right)^*\right.$$

$$\left. Q_0(x_0 - H_0\int \xi m(\xi)d\xi - \gamma_0) + v_0^*R_0v_0\right]$$

$$h_0(x_0,m) = 0 \tag{8.61}$$

$$g(x,x_0,m,v) = Ax + \Gamma x_0 + Bv + F\int \xi m(\xi)d\xi$$

$$\sigma(x) = \sigma \tag{8.62}$$

$$f(x,x_0,m,v) = \frac{1}{2}\left[\left(x - Hx_0 - \bar{H}\int \xi m(\xi)d\xi - \gamma\right)^*\right.$$

$$Q\left(x - Hx_0 - \bar{H}\int \xi m(\xi)d\xi - \gamma\right) + v^*Rv\bigg]$$

$$h(x,x_0,m) = 0. \tag{8.63}$$

We deduce easily

$$\hat{v}_0(x_0,m,p) = -R_0^{-1}B_0^*p \tag{8.64}$$

$$H_0(x_0,m,p) = \frac{1}{2}\left(x_0 - H_0\int \xi m(\xi)d\xi - \gamma_0\right)^* Q_0\left(x_0 - H_0\int \xi m(\xi)d\xi - \gamma_0\right)$$

$$+ p.\left(A_0x_0 + F_0\int \xi m(\xi)d\xi\right) - \frac{1}{2}p^*B_0R_0^{-1}B_0^*p \tag{8.65}$$

$$G_0(x_0,m,p) = A_0x_0 + F_0\int \xi m(\xi)d\xi - B_0R_0^{-1}B_0^*p \tag{8.66}$$

and

$$\hat{v}(x,x_0,m,q) = -R^{-1}B^*q \tag{8.67}$$

$$H(x,x_0,m,q) = \frac{1}{2}\left(x - Hx_0 - \bar{H}\int \xi m(\xi)d\xi - \gamma\right)^* Q\left(x - Hx_0 - \bar{H}\int \xi m(\xi)d\xi - \gamma\right)$$

$$+ q.\left(Ax + \Gamma x_0 + F\int \xi m(\xi)d\xi\right) - \frac{1}{2}q^*BR^{-1}B^*q \tag{8.68}$$

$$G(x,x_0,m,q) = Ax + \Gamma x_0 + F\int \xi m(\xi)d\xi - BR^{-1}B^*q. \tag{8.69}$$

We define

$$z(t) = \int \xi m(\xi,t)d\xi$$

and conjecture

$$u(x,t) = \frac{1}{2}x^*P(t)x + x^*r(t) + s(t) \tag{8.70}$$

$$K_l(x,t) = x^*K_l(t) + k_l(t).$$

We deduce that $P(t)$ is the solution of the Riccati equation

$$P' + PA + A^*P - PBR^{-1}B^*P + Q = 0$$

$$P(T) = 0. \qquad (8.71)$$

and the equations for $r(t)$ and $s(t)$

$$-dr = (A^* - P(t)BR^{-1}B^*)r(t)dt + [(P(t)F - Q\bar{H})z(t)$$
$$+ (P(t)\Gamma - QH)x_0(t) - Q\gamma]dt - \sum_l K_l(t)dw_{0l}(t)$$

$$r(T) = 0 \qquad (8.72)$$

$$-ds = \left[\text{tra}P(t) + r(t).(Fz(t) + \Gamma x_0(t)) - \frac{1}{2}r(t)^*BR^{-1}B^*r(t) \right.$$
$$\left. + \frac{1}{2}(Hx_0(t) + \bar{H}z(t) + \gamma)^*Q(Hx_0(t) + \bar{H}z(t) + \gamma) \right] dt$$
$$- \sum_l k_l(t)dw_{0l}(t). \qquad (8.73)$$

From the equation of $m(x,t)$ we get easily

$$\frac{dz}{dt} = (A - BR^{-1}B^*P(t) + F)z(t) + \Gamma x_0(t) - BR^{-1}B^*r(t)$$

$$z(0) = \bar{\omega} \qquad (8.74)$$

in which

$$\bar{\omega} = \int x\varpi(x)dx.$$

We proceed by computing

$$H_{0,x_0}(x_0, m, p) = Q_0\left(x_0 - H_0\int \xi m(\xi)d\xi - \gamma_0\right) + A_0^*p$$

$$\frac{\partial H_0}{\partial m}(x_0, m, p)(\xi) = \xi^*\left[-H_0^*Q_0(x_0 - H_0\int um(u)du - \gamma_0) + F_0^*p\right]$$

$$G_{x_0}(x,x_0,m,q) = \Gamma$$

$$\frac{\partial G}{\partial m}(x,x_0,m,q)(\xi) = F\xi$$

$$G_q(x,x_0,m,q) = -BR^{-1}B^*$$

$$H_{x_0}(x,x_0,m,q) = -H^*Q\left(x - Hx_0 - \bar{H}\int \xi m(\xi)d\xi - \gamma\right) + \Gamma^*q$$

$$\frac{\partial H}{\partial m}(x,x_0,m,q)(\xi) = \xi^*\left[-\bar{H}^*Q(x - Hx_0 - \bar{H}\int um(u)du - \gamma) + F^*q\right].$$

From the equations of the major player, we introduce the mean

$$\chi(t) = \int x\zeta(x,t)dx$$

and we postulate

$$\eta(x,t) = x^*\lambda(t) + \theta(t). \tag{8.75}$$

After some easy calculations, we obtain the following relations

$$dx_0 = (A_0 x_0(t) + F_0 z(t) - B_0 R_0^{-1} B_0^* p(t))dt + \sigma_0 dw_0(t)$$
$$x_0(0) = \xi_0 \tag{8.76}$$

$$-dp = [A_0^* p(t) + \Gamma^*\lambda(t) + Q_0(x_0(t) - H_0 z(t) - \gamma_0)$$
$$+ (\Gamma^* P(t) - H^* Q)\chi(t)]dt - \sum_{l=1}^{k_0} q_l(t)dw_{0l}(t)$$
$$p(T) = 0 \tag{8.77}$$

$$\frac{d\chi}{dt} = (A - BR^{-1}B^* P(t))\chi(t) - BR^{-1}B^*\lambda(t)$$
$$\chi(0) = 0 \tag{8.78}$$

$$-d\lambda = [(A^* - P(t)BR^{-1}B^* + F^*)\lambda(t) - H_0^*Q_0(x_0(t) - H_0z(t) - \gamma_0)$$

$$+ F_0^*p(t) + (F^*P(t) - \bar{H}^*Q)\chi(t)]dt - \sum_{l=1}^{k_0} \mu_l(t)dw_{0l}(t)$$

$$\lambda(T) = 0. \tag{8.79}$$

Therefore, we need to solve the system of (6.41)–(6.44) and (6.37) and (6.39). We obtain the optimal controls

$$\hat{v}(x,t) = -R^{-1}B^*(P(t)x + r(t))$$

$$\hat{v}_0(t) = -R_0^{-1}B_0^*p(t). \tag{8.80}$$

Note that $\theta(t)$ is given by

$$-d\theta = \lambda(t).(Fz(t) + \Gamma x_0(t) - BR^{-1}B^*r(t))dt - \sum_{l=1}^{k_0} v_l(t)dw_{0l}(t)$$

$$\theta(T) = 0. \tag{8.81}$$

In the framework of [31] we discard $\lambda(t)$ and $\chi(t)$. We get the system of four equations

$$dx_0 = (A_0x_0(t) + F_0z(t) - B_0R_0^{-1}B_0^*p(t))dt + \sigma_0dw_0(t) \tag{8.82}$$

$$x_0(0) = \xi_0$$

$$-dp = [A_0^*p(t) + Q_0(x_0(t) - H_0z(t) - \gamma_0)]dt - \sum_{l=1}^{k_0} q_l(t)dw_{0l}(t)$$

$$p(T) = 0 \tag{8.83}$$

$$\frac{dz}{dt} = (A - BR^{-1}B^*P(t) + F)z(t) + \Gamma x_0(t) - BR^{-1}B^*r(t)$$

$$z(0) = \bar{\varpi} \tag{8.84}$$

$$-dr = (A^* - P(t)BR^{-1}B^*)r(t)dt + [(P(t)F - Q\bar{H})z(t)$$

$$+ (P(t)\Gamma - QH)x_0(t) - Q\gamma]dt - \sum_l K_l(t)dw_{0l}(t)$$

$$r(T) = 0. \tag{8.85}$$

The case considered by Huang [19] is somewhat intermediary. It amounts to neglecting $\zeta(x,t)$. Since $\zeta(x,t)$ is the adjoint associated to the state $u(x,t)$ it means that we consider $u(x,t)$ as external in the m equation. But $m(t)$ is not external, it is influenced by $x_0(t)$. So we get five equations. In the LQ case it boils down to taking $\chi(t) = 0$ but not $\lambda(t)$.

We get the system

$$dx_0 = (A_0 x_0(t) + F_0 z(t) - B_0 R_0^{-1} B_0^* p(t)) dt + \sigma_0 dw_0(t)$$

$$x_0(0) = \xi_0 \tag{8.86}$$

$$-dp = [A_0^* p(t) + \Gamma^* \lambda(t) + Q_0(x_0(t) - H_0 z(t) - \gamma_0)] dt - \sum_{l=1}^{k_0} q_l(t) dw_{0l}(t)$$

$$p(T) = 0 \tag{8.87}$$

$$-d\lambda = [(A^* - P(t) B R^{-1} B^* + F^*) \lambda(t) - H_0^* Q_0(x_0(t)$$

$$- H_0 z(t) - \gamma_0) + F_0^* p(t)] dt - \sum_{l=1}^{k_0} \mu_l(t) dw_{0l}(t)$$

$$\lambda(T) = 0 \tag{8.88}$$

$$\frac{dz}{dt} = (A - B R^{-1} B^* P(t) + F) z(t) + \Gamma x_0(t) - B R^{-1} B^* r(t)$$

$$z(0) = \bar{\omega} \tag{8.89}$$

$$-dr = (A^* - P(t) B R^{-1} B^*) r(t) dt + [(P(t) F - Q \bar{H}) z(t)$$

$$+ (P(t) \Gamma - QH) x_0(t) - Q \gamma] dt - \sum_l K_l(t) dw_{0l}(t)$$

$$r(T) = 0. \tag{8.90}$$

Note that in the LQ case one can obtain the necessary conditions directly, without having to solve the general problem.

Chapter 9
Nash Differential Games with Mean Field Effect

9.1 Description of the Problem

The mean field game and mean field type control problems introduced in Chap. 2 are both control problems for a representative agent, with mean field terms influencing both the evolution and the objective functional of this agent. The terminology game comes from the fact that the optimal feedback of the representative agent can be used as an approximation for a Nash equilibrium of a large community of agents that are identical. In Sect. 8.2 we have shown that the theory extends to a multi-class of representative agents. However, each class still has its individual control problem.

A natural and important extension concerns the case of large coalitions competing with one another. This problem has been considerd in [7]. We present here the general situation. However, we use as interpretation the dual game concept, extending the considerations of Sect. 7.3. The interpretation as differential games among large coalitions remains to be done.

9.2 Mathematical Problem

We can generalize the pair of HJB-FP equations (3.11) considered in the mean field game to a system of N pairs of HJB-FP equations. Note that in this context N is fixed and will not tend to $+\infty$. We use the following notation

$$q = (q^1, \ldots, q^N), q^i \in \mathbb{R}^n$$
$$v = (v^1, \ldots, v^N), v^i \in \mathbb{R}^d$$

A. Bensoussan et al., *Mean Field Games and Mean Field Type Control Theory*,
SpringerBriefs in Mathematics, DOI 10.1007/978-1-4614-8508-7_9,
© Alain Bensoussan, Jens Frehse, Phillip Yam 2013

We consider functions

$$f^i(x,v) : \mathbb{R}^n \times \mathbb{R}^{dN} \to \mathbb{R}$$

$$g^i(x,v) : \mathbb{R}^n \times \mathbb{R}^{dN} \to \mathbb{R}^n$$

and define Lagrangians

$$L^i(x,v,q^i) = f^i(x,v) + q^i \cdot g^i(x,v) \tag{9.1}$$

We look for a Nash point for the Lagrangians in the controls $v^1, \ldots, v^N$. We write the system of equations

$$\frac{\partial L^i}{\partial v^i}(x,v,q) = 0 \tag{9.2}$$

Assuming that we can solve this system, we obtain functions $\hat{v}^i(x,q)$, which we call a Nash equilibrium for the Lagrangians. We note the vector of $\hat{v}^i(x,q)$, $\hat{v}(x,q)$. We next define the Hamiltonians

$$H^i(x,q) = L^i(x,\hat{v}(x,q),q^i) \tag{9.3}$$

and

$$G^i(x,q) = g^i(x,\hat{v}(x,q)). \tag{9.4}$$

Consider next probability densities $m^i(.)$ on $\mathbb{R}^n$. We set also

$$m(.) = (m^1(.),\ldots,m^N(.)).$$

These probablity densities are considered as elements of $L^1(\mathbb{R}^n)$.

We can now define the system of pairs of HJB-HP equations. We look for functions $u^i(x,t)$ and $m^i(x,t)$. We call the vector $u(x,t)$, $m(x,t)$. When we write $Du(x,t)$ we mean $(Du^1(x,t),\ldots,Du^N(x,t))$ so it is an $n \times N$ matrix. Define functions $f_0^i(x,m)$ and $h^i(x,m)$ defined on $\mathbb{R}^n \times L^1(\mathbb{R}^n)$. We set the system of pairs of PDEs

$$-\frac{\partial u^i}{\partial t} + Au^i = H^i(x,Du) + f_0^i(x,m^i(t))$$

$$u^i(x,T) = h^i(x,m^i(T)) \tag{9.5}$$

$$\frac{\partial m^i}{\partial t} + A^* m^i + \mathrm{div}\,(G^i(x,Du)m^i) = 0$$

$$m^i(x,0) = m_0^i(x) \tag{9.6}$$

which represents a generalization of the pair (3.11).

9.3 Interpretation

We can now interpret the system of pairs (9.5) and (9.6). The easiest way is to proceed as in Sect. 7.3. However, we need to assume that

$$f_0^i(x,m) = \frac{\partial \Phi^i(m)}{\partial m}(x) \tag{9.7}$$

$$h^i(x,m) = \frac{\partial \Psi^i(m)}{\partial m}(x). \tag{9.8}$$

Here m is an argument $\in L^1(R^n)$.

We consider N players. Each of them chooses a feedback control. So we have

$$v(.) = (v^1(.),\dots,v^N(,)).$$

Considering player i, we use the notation

$$v(.) = (v^i(.),\bar{v}^i(.))$$

where $\bar{v}^i(.)$ represents all the feedbacks except $v^i(.)$. To a vector of feedbacks $v(.)$ we associate the probabilities

$$p_{v(.)}^i(x,t) - p_{v^i(.),\bar{v}^i(.)}^i(x,t)$$

as the solution of

$$\frac{\partial p_{v(.)}^i}{\partial t} + A^* p_{v(.)}^i + \operatorname{div}\,(g^i(x,v(x))p_{v(.)}^i(x)) = 0$$

$$p_{v(.)}^i(x,0) = m_0^i(x). \tag{9.9}$$

We define the functional

$$J^i(v(.)) = \int_0^T \int_{R^n} p_{v(.)}^i(x,t)f^i(x,v(x))dxdt$$

$$+ \int_0^T \Phi^i(p_{v(.)}^i(t))dt + \Psi^i(p_{v(.)}^i(T)) \tag{9.10}$$

where $p_{v(.)}^i(t)$ means the function $p_{v(.)}^i(x,t)$. We need to compute the Gateaux differential of $J^i(v(.))$ with respect to $v^i(.)$, namely the quantity

$$\frac{d}{d\theta}J^i(v^i(.) + \theta\tilde{v}^i(.), \tilde{v}^i(.))|_{\theta=0} = \int_0^T \int_{\mathbb{R}^n} \tilde{m}^i(x,t)f^i(x,v(x))dxdt$$

$$+ \int_0^T \int_{\mathbb{R}^n} p^i_{v(.)}(x,t)\frac{\partial f^i(x,v(x))}{\partial v^i}\tilde{v}^i(x,t)dxdt$$

$$+ \int_0^T \int_{\mathbb{R}^n} f^i_0(x, p^i_{v(.)}(t))\tilde{m}^i(x,t)dxdt$$

$$+ \int_{\mathbb{R}^n} h^i(x, p^i_{v(.)}(T))\tilde{m}^i(x,T)dx \qquad (9.11)$$

where $\tilde{m}^i(x,t)$ is the solution of

$$\frac{\partial \tilde{m}^i}{\partial t} + A^*\tilde{m}^i + \mathrm{div}\,(g^i(x,v(x))\tilde{m}^i) + \mathrm{div}\,\left(\frac{\partial g^i(x,v(x))}{\partial v^i}\tilde{v}^i(x)p^i_{v(.)}(x,t)\right) = 0$$

$$p^i_{v(.)}(x,0) = m^i_0(x). \qquad (9.12)$$

One can then introduce the functions $u^i_{v(.)}(x,t)$ solutions of

$$-\frac{\partial u^i_{v(.)}}{\partial t} + Au^i_{v(.)} - g^i(x,v(x)) \cdot Du^i_{v(.)}(x) = f^i(x,v(x)) + f^i_0(x, p^i_{v(.)}(t))$$

$$u^i_{v(.)}(x,T) = h^i(x, p^i_{v(.)}(T)) \qquad (9.13)$$

and one can check easily that

$$\frac{d}{d\theta}J^i(v^i(.) + \theta\tilde{v}^i(.), \tilde{v}^i(.))|_{\theta=0} = \int_0^T \int_{\mathbb{R}^n} \frac{\partial L^i}{\partial v^i}(x,v(x), Du^i_{v(.)}(x))\tilde{v}^i(x)p^i_{v(.)}(x,t)dxdt.$$

$$(9.14)$$

A Nash point of functionals $J^i(v(.))$, $\hat{v}(.)$ must satisfy

$$\frac{\partial L^i}{\partial v^i}(x,\hat{v}(x), Du^i_{\hat{v}(.)}(x)) = 0.$$

It follows that if we set

$$u^i(x,t) = u^i_{\hat{v}(.)}(x,t), \quad m^i(x,t) = p^i_{\hat{v}(.)}(x,t)$$

$$\hat{v}(x,t) = \hat{v}(x, Du(x,t))$$

then the system of pairs u^i, m^i is a solution of (9.5) and (9.6). Hence $\hat{v}(.)$ is a Nash equilibrium for problems (9.9) and (9.10).

We can give a probabilistic interpretation to problems (9.9) and (9.10). Consider N independent standard Wiener processes $w^i(t)$ and N independent random variables x_0^i, with a probability density of m_0^i. The random variables are also independent of the Wiener processes. For a vector of feedbacks the representative agents have states $x^i(.) = x_{v(.)}^i(.)$ solutions of the equations

$$dx^i = g^i(x^i(t), v(x^i(t)))dt + \sigma(x^i(t))dw^i(t)$$
$$x^i(0) = x_0^i. \tag{9.15}$$

It is clear that the probability density of $x^i(t)$ is $p_{v(.)}^i(t)$. We note

$$P_{x^i(t)} = p_{v(.)}^i(t)$$

and we have

$$J^i(v(.)) = E \int_0^T f^i(x^i(t), v(x^i(t)))dt + \int_0^T \Phi^i(P_{x^i(t)})dt + \Psi^i(P_{x^i(T)}). \tag{9.16}$$

In this problem, each player sees the feedbacks of his or her opponents acting on his or her own trajectory. All the trajectories are independent.

9.4 Another Interpretation

We can give another interpretation related to the mean field game interpretation for a single player; see Chap. 2. For given functions $m^i(t)$, deterministic, to fix the ideas in $C([0, T]; \mathbb{R}^n)$ we consider the following Nash equilibrium problem. We have state equations, as in (9.15)

$$dx^i = g^i(x^i(t), v(x^i(t)))dt + \sigma(x^i(t))dw^i(t)$$
$$x^i(0) = x_0^i \tag{9.17}$$

controlled by feedbacks $v(.) = (v^1(.), \ldots, v^N(.))$ and payoffs

$$J^i(v(.), m^i(.)) = E \int_0^T [f^i(x^i(t), v(x^i(t))) + f_0^i(x^i(t), m^i(t))]dt$$
$$+ Eh^i(x^i(T), m^i(T)) \tag{9.18}$$

We look for a Nash equilibrium for problem (9.17) and (9.18). By classical methodology we obtain the system of HJB equations

$$-\frac{\partial u^i}{\partial t} + Au^i = H^i(x, Du) + f_0^i(x, m^i(t))$$
$$u^i(x, T) = h^i(x, m^i(T)) \tag{9.19}$$

with optimal feedbacks

$$\hat{v}(x,t) = \hat{v}(x, Du(x,t)).$$

If we use these optimal feedbacks in the state equations (9.17) we obtain the trajectories of the Nash equilibrium

$$d\hat{x}^i = g^i(\hat{x}^i(t), \hat{v}(\hat{x}^i(t)))dt + \sigma(\hat{x}^i(t))dw^i(t)$$

$$\hat{x}^i(0) = x_0^i. \tag{9.20}$$

We now request that the functions $m^i(t)$ represent the probability densities of the trajectories $\hat{x}^i(t)$. Clearly the functions $m^i(t)$ are given by the FP equations

$$\frac{\partial m^i}{\partial t} + A^* m^i + \mathrm{div}\,(G^i(x, Du)m^i) = 0$$

$$m^i(x,0) = m_0^i(x) \tag{9.21}$$

and

$$J^i(\hat{v}(.), m^i(.)) = \int_{\mathbb{R}^n} u^i(x,0)m_0^i(x)dx. \tag{9.22}$$

9.5 Generalization

We can introduce more general problems than (9.5) and (9.6). We can write

$$-\frac{\partial u^i}{\partial t} + Au^i = H^i(x, m, Du)$$

$$u^i(x,T) = h^i(x, m(T)) \tag{9.23}$$

$$\frac{\partial m^i}{\partial t} + A^* m^i + \mathrm{div}\,(G^i(x, m, Du)m^i) = 0$$

$$m^i(x,0) = m_0^i(x) \tag{9.24}$$

in which $m = (m^1, \ldots, m^N)$ and the functions H^i, G^i depend on the full vector m. The interpretation is much more elaborate.

9.6 Approximate Nash Equilibrium for Large Communities

We now extend the theory developed in Sect. 5.4. We want to associate to problems (9.5) and (9.6) a differential game for N communities, composed of very large numbers of agents. We denote the agents by the index i, j where $i = 1, \ldots, N$ and $j = 1, \ldots, M$. The number M will tend to $+\infty$. Each player i, j chooses a feedback $v^{i,j}(x)$, $x \in \mathbb{R}^n$. The state of player i, j is denoted by $x^{i,j}(t) \in \mathbb{R}^n$. We consider independent standard Wiener processes $w^{i,j}(t)$ and independent replicas $x_0^{i,j}$ of the random variable x_0^i, with the probability density m_0^i. They are independent of the Wiener processes. We denote

$$v^j(.) = (v^{1,j}(.), \ldots, v^{N,j}(.)).$$

The trajectory of the state $x^{i,j}$ is defined by the equation

$$dx^{i,j} = g^i(x^{i,j}, v^j(x^{i,j}))dt + \sigma(x^{i,j})dw^{i,j}$$
$$x^{i,j}(0) = x_0^{i,j}. \tag{9.25}$$

The trajectories are independent. The player i, j trajectory is influenced by the feedbacks $v^{k,j}(x)$, $k \neq i$, acting on his own state. When we focus on player i we use the notation

$$v^j(.) = (v^{i,j}(.), \bar{v}^{i,j}(.))$$

in which $\bar{v}^{i,j}(.)$ represents all feedbacks $v^{k,j}(x)$, $k \neq i$. The notation $v(.)$ represents all feedbacks.

We now define the objective functional of player i, j by

$$\mathcal{J}^{i,j}(v(.)) = E \int_0^T \left[f^i(x^{i,j}(t), v^j(x^{i,j}(t))) \right.$$

$$\left. + f_0^i\left(x^{i,j}(t), \frac{1}{M-1}\sum_{l=1\neq j}^{M} \delta_{x^{i,l}(t)}\right) \right] dt + Eh^i\left(x^{i,j}(T), \frac{1}{M-1}\sum_{l=1\neq j}^{M} \delta_{x^{i,l}(T)}\right).$$

$$\tag{9.26}$$

We look for a Nash equilibrium. Consider next the system of pairs of HJB-FP equations (9.5) and (9.6) and the feedback $\hat{v}(x)$. We want to show that the feedback

$$\hat{v}^{i,j}(.) = \hat{v}^i(.)$$

is an approximate Nash equilibrium. If we use this feedback in the state equation (9.25) we get

$$d\hat{x}^{i,j} = g^i(\hat{x}^{i,j}, \hat{v}(\hat{x}^{i,j}))dt + \sigma(\hat{x}^{i,j})dw^{i,j}$$

$$\hat{x}^{i,j}(0) = x_0^{i,j}$$

and the trajectories $\hat{x}^{i,j}$ become independent replicas of $\hat{x}^i$ solution of

$$d\hat{x}^i = g^i(\hat{x}^i, \hat{v}(\hat{x}^i))dt + \sigma(\hat{x}^i)dw^i$$

$$\hat{x}^i(0) = x_0^i.$$

The probability density of $\hat{x}^i(t)$ is $m^i(t)$. Therefore,

$$\mathcal{J}^{i,j}(\hat{v}(.)) - J^i(\hat{v}(.)) = E \int_0^T \left[f_0^i\left(\hat{x}^{i,j}(t), \frac{1}{M-1} \sum_{l=1\neq j}^{M} \delta_{\hat{x}^{i,l}(t)} \right) - f_0^i(\hat{x}^i(t), m^i(t)) \right] dt$$

$$+ E\left[h^i\left(\hat{x}^{i,j}(T), \frac{1}{M-1} \sum_{l=1\neq j}^{M} \delta_{\hat{x}^{i,l}(T)} \right) - h^i(\hat{x}^i(T), m^i(T)) \right]$$

and also

$$\mathcal{J}^{i,j}(\hat{v}(.)) - J^i(\hat{v}(.), m^i(.)) = E \int_0^T \left[f_0^i\left(\hat{x}^{i,j}(t), \frac{1}{M-1} \sum_{l=1\neq j}^{M} \delta_{\hat{x}^{i,l}(t)} \right) \right.$$

$$\left. - f_0^i(\hat{x}^{i,j}(t), m^i(t)) \right] dt$$

$$+ E\left[h^i\left(\hat{x}^{i,j}(T), \frac{1}{M-1} \sum_{l=1\neq j}^{M} \delta_{\hat{x}^{i,l}(T)} \right) - h^i(\hat{x}^{i,j}(T), m^i(T)) \right].$$

For fixed i, the variables $\hat{x}^{i,l}(t), l = 1, \ldots, M$ are independent and distributed with the density $m^i(t)$. By arguments already used, see Sect. 5.4, the random measure in $\mathbb{R}^n$ converges a.s. towards $m^i(x,t)dx$ for the topology of weak* convergence of measures on $\mathbb{R}^n$. Provided the functionals $f_0^i(x,m)$ and $h^i(x,m)$ are continuous in m for the topology of weak* convergence of measures on $\mathbb{R}^n$, for any fixed x, and provided the Lebesgue's theorem can be used, we can assert that

$$\mathcal{J}^{i,j}(\hat{v}(.)) - J^i(\hat{v}(.), m^i(.)) \to 0, \text{ as } M \to +\infty.$$

We now focus on player $1,1$ to fix the ideas. Suppose he or she uses a feedback $v^{1,1}(x) \neq \hat{v}^{1,1}(x)$, and the other players use $\hat{v}^{i,j}(x) = \hat{v}^i(x), \forall i \geq 2, \forall j$ or $\forall i, \forall j \geq 2$. We set $v^1(x) = v^{1,1}(x)$. Call this set of controls $\bar{v}(.)$. By abuse of notation, we also call

$$\bar{v}(.) = (v^1(.), \hat{v}^2(.), \ldots, \hat{v}^N(.)) = (v^1(.), \bar{v}^1(.)).$$

The corresponding trajectories are denoted by $y^{1,j}(t)$ solutions of

$$dy^{1,1} = g^1(y^{1,1}, v^1(y^{1,1}), \bar{v}^1(y^{1,1}))dt + \sigma(y^{1,1})dw^{1,1}$$
$$y^{1,1}(0) = x_0^{1,1} \tag{9.27}$$

and $y^{1,j} = \hat{x}^{1,j}$ for $j \geq 2$.

We can then compute

$$J^{1,1}(\bar{v}(.)) = E \int_0^T f^1(y^{1,1}(t), v^1(y^{1,1}), \bar{v}^1(y^{1,1}))dt$$

$$+ E \int_0^T f_0^1 \left(y^{1,1}(t), \frac{1}{M-1} \sum_{l=2}^M \hat{x}^{1,l}(t) \right) dt + E h^1 \left(y^{1,1}(T), \frac{1}{M-1} \sum_{l=2}^M \delta_{\hat{x}^{1,l}(T)} \right)$$

$$= E \int_0^T f^1(y^{1,1}(t), v^1(y^{1,1}), \bar{v}^1(y^{1,1}))dt$$

$$+ E \int_0^T f_0^1(y^{1,1}(t), m^1(t))dt + E h^1(y^{1,1}(T), m^1(T))$$

$$+ E \int_0^T f_0^1 \left(y^{1,1}(t), \frac{1}{M-1} \sum_{l=2}^M \hat{x}^{1,l}(t) \right) dt - E \int_0^T f_0^1(y^{1,1}(t), m^1(t))dt$$

$$+ E h^1 \left(y^{1,1}(T), \frac{1}{M-1} \sum_{l=2}^M \delta_{\hat{x}^{1,l}(T)} \right) - E h^1(y^{1,1}(T), m^1(T))$$

$$\geq \int_{\mathbb{R}^n} u^1(x,0)m_0^1(x)dx$$

$$+ E \int_0^T f_0^1 \left(y^{1,1}(t), \frac{1}{M-1} \sum_{l=2}^M \hat{x}^{1,l}(t) \right) dt - E \int_0^T f_0^1(y^{1,1}(t), m^1(t))dt$$

$$+ E h^1 \left(y^{1,1}(T), \frac{1}{M-1} \sum_{l=2}^M \delta_{\hat{x}^{1,l}(T)} \right) - E h^1(y^{1,1}(T), m^1(T)). \tag{9.28}$$

Recalling that

$$\int_{\mathbb{R}^n} u^1(x,0)m_0^1(x)dx = J^1(\hat{v}(.), m^1(.))$$

and using previous convergence arguments, we obtain

$$J^{1,1}(\bar{v}(.)) \geq J^1(\hat{v}(.), m^1(.)) - O\left(\frac{1}{\sqrt{M}}\right)$$

and this concludes the approximate Nash equilibrium property.

Chapter 10
Analytic Techniques

10.1 General Set-Up

10.1.1 Assumptions

We consider here the system

$$-\frac{\partial u^i}{\partial t} + Au^i = H^i(x, Du) + f_0^i(x, m(t))$$

$$u^i(x, T) = h^i(x, m(T)) \tag{10.1}$$

$$\frac{\partial m^i}{\partial t} + A^* m^i + \mathrm{div}\,(G^i(x, Du)m^i) = 0$$

$$m^i(x, 0) = m_0^i(x) \tag{10.2}$$

which we consider as a system of PDEs. We call u, m the vectors with components u^i, m^i. We want to give an existence and a regularity result. We recall

$$A\varphi(x) = -\sum_{k,l=1}^{n} a_{k,l}(x) \frac{\partial^2 \varphi}{\partial x_k \partial x_l}(x). \tag{10.3}$$

$$A^*\varphi(x) = -\sum_{k,l=1}^{n} \frac{\partial^2}{\partial x_k \partial x_l}(a_{kl}(x)\varphi(x)). \tag{10.4}$$

in the mean field problem $x \in \mathbb{R}^n$. To simplify the analytic treatment, we take $x \in \mathcal{O}$ as a Lipschitz bounded domain with a boundary denoted by $\partial\mathcal{O}$. We assume

A. Bensoussan et al., *Mean Field Games and Mean Field Type Control Theory*,
SpringerBriefs in Mathematics, DOI 10.1007/978-1-4614-8508-7__10,
© Alain Bensoussan, Jens Frehse, Phillip Yam 2013

$$a_{k,l} = a_{l,k} \in W^{1,\infty}(\mathcal{O})$$

$$\sum_{k,l=1}^{n} a_{k,l}(x)\xi_k\xi_l \geq \alpha|\xi|^2, \forall \xi \in \mathbb{R}^n, \forall x \in \mathcal{O}, \alpha > 0. \tag{10.5}$$

We need to specify boundary conditions for problems (10.1) and (10.2). In this presentation, we shall use Neumann boundary conditions, but Dirichlet boundary conditions are possible and in fact simpler. If $x \in \partial\mathcal{O}$, we call $v(x)$ the outward unit normal at x. The Neumann boundary condition for (10.1) reads

$$\sum_{k,l=1}^{n} v_k(x)a_{k,l}(x)\frac{\partial u^i}{\partial x_l}(x) = 0, \forall x \in \partial\mathcal{O}, i = 1,\dots,N \tag{10.6}$$

and the Neumann boundary condition for (10.2) reads

$$\sum_{k,l=1}^{n} v_l(x)\left[\frac{\partial}{\partial x_k}(a_{k,l}(x)m^i(x)) - G_l^i(x, Du(x))m^i(x)\right] = 0, \forall x \in \partial\mathcal{O}, i = 1,\dots,N. \tag{10.7}$$

To simplify notation we set

$$L^{\infty}(L^p) = L^{\infty}(0,T;L^p(\mathcal{O},\mathbb{R}^N)), 1 \leq p \leq \infty$$

$$C(L^2) = C([0,T];L^2(\mathcal{O},\mathbb{R}^N)).$$

We shall also need the Sobolev spaces

$$L^2(W^{1,2}) = L^2(0,T;W^{1,2}(\mathcal{O},\mathbb{R}^N))$$

$$L^2((W^{1,2})^*) = L^2(0,T;(W^{1,2})^*(\mathcal{O},\mathbb{R}^N)).$$

We may also use the short notation for the components of the vector in $\mathbb{R}^N$.
We next make the assumptions

$$H^i(x,t;q), G^i(x,t;q) : \mathbb{R}^n \times \mathbb{R} \times \mathbb{R}^{nN} \text{ are measurable, continuous in } q \tag{10.8}$$

$$|H^i(x,t;q)| + |G^i(x,t;q)|^2 \leq K|q|^2 + K \tag{10.9}$$

and

$$f_0^i(x,t,m),: \mathbb{R}^n \times \mathbb{R} \times L^1(\mathcal{O},\mathbb{R}^N)), \text{measurable, continuous in } m \tag{10.10}$$

$$|f_0^i(x,t,m)| \leq k_0(\|m\|)$$

where

$$||m|| = ||m||_{L^1(\mathcal{O}, \mathbb{R}^N)}$$

and k_0 is bounded on bounded sets. Also

$$h^i(x,m),: \mathbb{R}^n \times L^1(\mathcal{O}, \mathbb{R}^N)), \text{measurable, continuous in } m$$

$$|h^i(x,m)| \le k_0(||m||)$$

$$m_0 \in L^2(\mathcal{O}, \mathbb{R}^N). \tag{10.11}$$

10.1.2 Weak Formulation

We now give a weak formulation of problems (10.1) and (10.2) . We look for a pair u, m such that

$$u, m \in L^2(W^{1,2}) \cap L^\infty(L^2) \tag{10.12}$$

$$m \in C(L^2), \ m^i G^i(., Du) \in L^2(L^2)$$

and u, m satisfy

$$\int_0^T (u^i, \dot{\varphi}^i)_{L^2} dt + \sum_{k,l=1}^n \int_0^T (D_l u^i, D_k(a_{kl}\varphi^i))_{L^2} dt = (h^i(., m(T)), \varphi^i(T))_{L^2}$$

$$+ \int_0^T (H^i(., Du) + f_0^i(., m(t)), \varphi^i)_{L^2} dt, \ \forall i = 1, \dots, N \quad (10.13)$$

for any test function $\varphi^i \in L^2(W^{1,2}) \cap L^\infty(L^\infty), \dot{\varphi}^i \in L^2((W^{1,2})^*)$ such that $\varphi^i(t)$ vanishes in a neighborhood of $t = 0$. Note that $\varphi^i \in C(L^2)$.
    Similarly,

$$-\int_0^T (m^i, \dot{\varphi}^i)_{L^2} dt + \sum_{k,l=1}^n \int_0^T (D_l \varphi^i, D_k(a_{kl} m^i))_{L^2} dt$$

$$= (m_0, \varphi^i(0))_{L^2} + \int_0^T (m^i G^i(., Du), D\varphi^i)_{L^2} dt, \ \forall i = 1, \dots, N \quad (10.14)$$

for any test function $\varphi^i \in L^2(W^{1,2}) \cap L^\infty(L^\infty), \dot{\varphi}^i \in L^2((W^{1,2})^*)$, such that $\varphi^i(t)$ vanishes in a neighborhood of $t = T$. We have denoted

$$(\varphi, \psi)_{L^2} = \int_{\mathcal{O}} \varphi(x)\psi(x)dx$$

and D_k stands for $\frac{\partial}{\partial x_k}$.

10.2   A Priori Estimates for u

10.2.1 L^∞ Estimate for u

We assume here the structure

$$|H^i(x,q)| \leq K|q^i||q| + K \qquad (10.15)$$

Proposition 7. *We assume (10.5), (10.10), (10.11), (10.15). Consider a weak solution of (10.13) with $m \in L^\infty(L^1)$, ≥ 0. Then*

$$||u||_{L^\infty(L^\infty)} \leq K_0 \qquad (10.16)$$

where K_o depends only on the $L^\infty(L^1)$ bound of m and of the various constants.

Proof. We note that $f_0^i(x,m(t)), h^i(x,m(T))$ are bounded by a constant. So, in fact, we have

$$\left| -\frac{\partial u^i}{\partial t} + Au^i \right| \leq K|Du^i||Du| + K$$

$$|u^i(x,T)| \leq K$$

Then there exists a function $\sigma(x,t)$ such that

$$||\sigma||_{L^\infty(L^\infty)} \leq 1$$

and

$$-\frac{\partial u^i}{\partial t} + Au^i = \sigma(K|Du^i||Du| + K).$$

Define

$$\tilde{G}^i(x,q) = K\sigma \frac{Du^i}{|Du^i|}|q|, \text{ if } Du^i \neq 0$$

$$= 0, \text{ if } Du^i = 0$$

then u^i satisfies

$$-\frac{\partial u^i}{\partial t} + Au^i = \tilde{G}^i(x,Du).Du^i + K\sigma.$$

The probabilistic interpretation, or the maximum principle, shows immediately that

$$|u(x_0,t_0)| \leq \max(\sup|u(x,T)|, KT)$$

which proves the result. $\square$

10.2.2 $L^2(W^{1,2})$ *Estimate for* u

Proposition 8. *We make the assumptions of Proposition 7, that a weak solution of (10.13) satisfies*

$$\|u\|_{L^2(W^{1,2})} \leq K_0 \tag{10.17}$$

where K_0 depends only on the $L^\infty(L^1)$ bound of m and of the various constants.

Proof. The proof is given in [3] and is not reproduced here. It relies on using the following test functions

$$\varphi^i = (\exp(\lambda u^i) - \exp(-\lambda u^i)) \exp\left[\frac{\gamma}{\alpha} \sum_{j=1}^{N} (\exp(\lambda u^j) - \exp(-\lambda u^j))\right]$$

with parameters λ, γ sufficiently large. More precisely, we use iterated exponentials related to φ^i (see details the reference above). In the reference, however Dirichlet conditions are assumed. But the proof carries over to Neumann boundary conditions. $\qquad\square$

10.2.3 C^α *Estimate for* u

We need to consider here, that the operator A is written in divergence form

$$A\varphi(x) = -D_k(a_{k,l}(x)D_l\varphi(x))$$

which can be done since $a_{k,l}(x)$ are Lipschitz continuous, with a modification of the Hamiltonian, which does not change the assumptions.

One makes use of the Green's function $\Gamma_{x_0,t_0}(x,t)$ solution of the backward equation, $t < t_0$,

$$-\frac{\partial \Gamma}{\partial t} + A\Gamma = 0$$

$$\Gamma_{x_0,t_0}(x,t_0) = \delta(x - x_0).$$

One can then prove the following estimate.

Proposition 9. *We make the assumptions of Proposition 7, and $\partial\mathcal{O}$ smooth; then one has the estimate*

$$\int_{t_0}^{T} \int_{\mathcal{O}} |Du|^2 \Gamma_{x_0,t_0}(x,t)dxdt \leq C \tag{10.18}$$

where C depends only on the $L^\infty(L^1)$ bound of m, of the various constants and of the domain.

Proof. In reference [3], the Dirichlet problem is considered. It suffices to test with $\varphi^i \Gamma_{x_0,t_0}$, after extending the solution u^i by 0, outside the domain $\mathcal{O}$. This is not valid for Neumann boundary conditions. Because the boundary is smooth, we can use the following procedure. First, the problem is transformed locally to the half-space $x_n > 0$. In a portion $U \cap \{x_n > 0\}$ the function u is extended by (a change of coordinates is necessary)

$$\tilde{u} = u \text{ on } U \cap \{x_n > 0\}$$

$$\tilde{u}(x_1,\ldots,x_{n-1},x_n,t) = u(x_1,\ldots,x_{n-1},-x_n,t),\ x_n < 0$$

If we extend the coefficients

$$a_{k,l}(x_1,\ldots,x_n) = a_{k,l}(x_1,\ldots,-x_n),\ x_n < 0,$$

$$\text{if } l \neq n, k \neq n \text{ nor } k = l = n$$

$$a_{n,l}(x_1,\ldots,x_n) = -a_{n,l}(x_1,\ldots,-x_n),\ l \neq n$$
$$a_{k,n}(x_1,\ldots,x_n) = -a_{k,n}(x_1,\ldots,-x_n),\ k \neq n$$

The Hamiltonian is extended as follows

$$H^i(x_1,\ldots,x_n,q^1,\ldots,q^{n-1},q^n) = H^i(x_1,\ldots,-x_n,q^1,\ldots,q^{n-1},-q^n)$$

and

$$f_0^i(x_1,\ldots,x_n,m(t)) = f_0^i(x_1,\ldots,-x_n,m(t))$$

The extended elliptic operator is discontinuous at the boundary, but still uniformly elliptic. Since it is in divergence form, it is valid. The Neumann condition holds on both sides of $x_n = 0$. The extended solution solves a parabolic problem, for which we need to consider only interior estimates. So if the new domain is denoted $\tilde{\mathcal{O}}$, we consider the Green's function with respect to $\tilde{\mathcal{O}}$ or a cube $\supset\supset \tilde{\mathcal{O}}$. One then uses a test function $\varphi^i \Gamma_{x_0,t_0} \tau^2$, where $\tau \geq 0$, is a localization function with compact support in $\tilde{\mathcal{O}}$. This yields the desired estimate. $\square$

From the preceding result, one can derive the C^α a priori estimate

Proposition 10. *With the assumptions of Proposition 9, one has the estimate*

$$[u]_{C^\alpha} \leq C, \tag{10.19}$$

where C depends only on the $L^\infty(L^1)$ bound of m, of the various constants and of the domain.

Proof. Again, we do not detail the proof and refer to [3, 6]. The idea is to use a Campanato test

$$\sup_R \left\{ R^{-n-2-2\alpha} \int_{Q_R} |u - \bar{u}_R|^2 dxdt \,|\, 0 < R < R_1, Q_R \subset \mathcal{O} \times (0,T) \right\} \leq C, \quad (10.20)$$

where Q_R is a parabolic cylinder of size R, around any point of the domain $\mathcal{O} \times (0,T)$, the sup is taken over all points and over all sizes. The quantity $\bar{u}_R$ is the mean value of u over Q_R. Alternatively, $\bar{u}_R$ can also be chosen to be the maen value over $Q_{mR} - Q_R$, m fixed.

The bound C is as indicated in the statement of the proposition. From this property, it results that u is Hölder continuous on domains Q such that $Q \subset\subset \mathcal{O} \times (0,T)$, with

$$[u]_{C^\alpha(Q)} \leq C_Q.$$

Due to the possibility of extending the solution across the boundary $\partial\mathcal{O} \times (0,T)$ and $\mathcal{O} \times \{0\}$, we can, in fact, state

$$[u]_{C^\alpha(\overline{\mathcal{O} \times (0,T)})} \leq C$$

with a bound as indicated in the statement of the proposition.

In our setting, the Campanato criterium is established, proving first the estimate

$$\int_{Q_R} |Du|^2 dxdt \leq CR^{n+2\alpha} \qquad (10.21)$$

which implies, via Poincaré's inequality,

$$\int_{t_0-R^2}^{t_0} \int_{B_R} \left| u - \fint_{B_R} u(x,t)dx \right|^2 dxdt \leq CR^{n+2+2\alpha},$$

where $B_R = B_R(x_0)$ is the ball of radius R and center x_0. From this and the equation, one can also estimate the differences of mean values

$$\left| \fint_{B_R} u(x,t_1)dx - \fint_{B_R} u(x,t_2)dx \right| \leq CR^{2\alpha}.$$

So the crucial point is to establish the Morrey inequality (10.21). The standard technique is to obtain a "hole-filling inequality." In the elliptic case, such inequalities are of the form

$$\int_{B_R} |Du|^2 G dx \leq C \int_{B_{2R}-B_R} |Du|^2 G dx + C R^{2\alpha},$$

where $G = G(.;x_0)$ is the fundamental solution of the underlying elliptic operator.

The parabolic analogue is much more complex since one has to deal with fundamental solutions at different time levels. The corresponding inequality reads

$$\int_{Q_R} |Du|^2 \Gamma dx dt \leq K(\varepsilon) \int_{Q_R - Q_{\frac{R}{2}}} |Du|^2 \Gamma dx dt$$

$$+ \delta(\varepsilon) R^{-n} \int_{Q_R - Q_{\frac{R}{2}}} |Du|^2 dx dt + C R^{\beta},$$

where Γ is the fundamental solution of the parabolic operator with singularity at (x_0, t_0). If we discard the term in $\delta(\varepsilon)$, then we can use the standard hole-filling procedure which implies

$$\int_{Q_{\frac{R}{2}}} |Du|^2 \Gamma dx dt \leq \frac{K(\varepsilon)}{1+K(\varepsilon)} \int_{Q_{4R}} |Du|^2 \Gamma dx dt + K R^{\beta}.$$

Then an iteration argument applied to $R = 2^{-k}$ yields

$$\int_{Q_R} |Du|^2 \Gamma dx dt \leq C R^{2\alpha}.$$

To deal with the term in $\delta(\varepsilon)$, one uses the fact that

$$K(\varepsilon) \sim \frac{1}{\varepsilon}, \quad \frac{\delta(\varepsilon)}{1+K(\varepsilon)} \sim \varepsilon^2.$$

One then uses a supremum argument, which covers the term in $\delta(\varepsilon)$ and takes into account that $\Gamma \geq c_0 R^n$ on $Q_{2R} - Q_R$, but not necessarily on Q_R(this is different from the elliptic case). One derives (10.21). $\square$

10.2.4 $L^p(W^{2,p})$ Estimate for u

We show here how all the previous a priori estimates on u, including the C^α estimate allow us to obtain estimates in the space $L^p(W^{2,p}), 2 \leq p < \infty$, provided we assume the following regularity in the final condition

$$\|h^i(.,m)\|_{L^\infty(W^{2,p})} \leq k_0(\|m\|) \tag{10.22}$$

$$\sum_{k,l=1}^{n} v_k(x) a_{k,l}(x) \frac{\partial h^i}{\partial x_l}(x,t,m) = 0, \forall x \in \partial \mathcal{O}, \forall m, \forall t\, i = 1,\ldots,N. \qquad (10.23)$$

We have the following

Proposition 11. *We make the assumptions of Proposition 9 and (10.22) and (10.23). Then, one has the estimate*

$$\|u\|_{L^p(W^{2,p})} \leq C, \ \left\|\frac{\partial u}{\partial t}\right\|_{L^p(L^p)} \leq C, \, p < \infty, \qquad (10.24)$$

where C depends only on the $L^\infty(L^1)$ bound of m, of the various constants and of the domain.

Proof. The assumptions (10.22) and (10.23) allow to reduce the final condition to

$$u^i(x,T) = 0$$

simply by considering the difference $u^i(x,t) - h^i(x,m(T))$. The second property (10.24) is a consequence of the first one. So it remains to prove the first one with 0 final condition. We shall use linear parabolic L^p theory. A delicate point in the discussion concerns the compatibility conditions for the boundary data on $(t = T) \times \partial \mathcal{O}$. We refer to [28], Theorem 5.3, p. 320. In our case, the compatibility condition is satisfied, thanks to (10.23) and the continuity of u, proved in the preceding section. A convenient reference for the $L^2(W^{2,p})$ property in the case of Dirichlet boundary conditions is the paper of Schlag, [33]. We believe that his technique can be adapted to the Neumann case. Nevertheless, we proceed with local estimates and local charts to treat the boundary.

We first prove a local estimate, namely if $\overline{\mathcal{O}_0} \subset \mathcal{O}$, then

$$\|u\|_{L^p(W^{2,p}(\mathcal{O}_0))} \leq C_0. \qquad (10.25)$$

Let $x_0 \in \mathcal{O}$ and $t_0 \in (0,T)$. We consider a ball $B_{2R} \subset \mathcal{O}$. The radius R will be small, but will not tend to 0. We set

$$Q_R = B_R \times ((t_0 - R^2)^+, t_0)$$

and we consider a test function $\tau(x,t)$ that is equal to 1 on Q_R, with support included in Q_{2R}, which is Lipschitz, with second derivatives in x bounded.

We note that

$$\left| -\frac{\partial u^i}{\partial t} + Au^i \right| \leq K|Du|^2 + K.$$

Let u_R^i be the mean value of u over Q_R. It is easy to check the following localized inequality

$$\left| -\frac{\partial}{\partial t}((u^i - u_R^i)\tau^2) + A((u^i - u_R^i)\tau^2) \right| \le K|D((u - u_R)\tau)|^2 + K(\tau)$$

in which K does not depend on τ, but the second constant depends on τ, its derivatives, and the L^∞ bound on u. We then use the standard $L^p(W^{2,p})$ theory of parabolic equations with Dirichlet conditions, to claim the estimate

$$\int_0^T \int_{\mathcal{O}} \left| \frac{\partial}{\partial t}((u - u_R)\tau^2) \right|^p dxdt + \int_0^T \int_{\mathcal{O}} |D^2((u - u_R)\tau^2)|^p dxdt$$

$$\le K \int_0^T \int_{\mathcal{O}} |D((u - u_R)\tau)|^{2p} dxdt + K(\tau). \tag{10.26}$$

We skip details, which are easy. Setting $w = (u - u_R)\tau$, which vanishes on $\partial \mathcal{O}$, we may write, by integration by parts

$$\int_0^T \int_{\mathcal{O}} |D_k w^i|^{2p} dxdt = -(2p - 1) \int_0^T \int_{\mathcal{O}} w^i D_k^2 w^i |D_k w^i|^{2p-2} dxdt$$

hence

$$\int_0^T \int_{\mathcal{O}} |D_k w^i|^{2p} dxdt \le K_p \int_0^T \int_{\mathcal{O}} |w^i D_k^2 w^i|^p dxdt$$

which yields easily

$$\int_0^T \int_{\mathcal{O}} |D_k w^i|^{2p} dxdt \le K_p \int_0^T \int_{\mathcal{O}} |u^i - u_R^i|^p |D_k^2(w^i\tau)|^p dxdt + K(\tau).$$

Thanks to the fact that the functions u^i are C^α, we can choose R sufficiently small so that this inequality becomes

$$\int_0^T \int_{\mathcal{O}} |D_k w^i|^{2p} dxdt \le \varepsilon \int_0^T \int_{\mathcal{O}} |D_k^2(w^i\tau)|^p dxdt + K(\tau)$$

in which ε can be chosen small. Adding terms we get

$$\int_0^T \int_{\mathcal{O}} |D((u - u_R)\tau)|^{2p} dxdt \le \varepsilon \int_0^T \int_{\mathcal{O}} |D^2((u - u_R)\tau^2)|^p dxdt + K(\tau)$$

which used in (10.25) yields

$$\int_0^T \int_{\mathcal{O}} \left| \frac{\partial}{\partial t}((u - u_R)\tau^2) \right|^p dxdt + \int_0^T \int_{\mathcal{O}} |D^2((u - u_R)\tau^2)|^p dxdt \le K(\tau).$$

By combining a finite number of small domains, we obtain the property (10.25). To obtain the estimate up to the boundary, we need to use a system of local charts. To give all details is too technical and too long. We just explain the generic problems. Suppose we have a domain $\mathcal{O}'$ in $x_n \geq 0$, its boundary Γ contains a part $\Gamma' \subset \{x_n = 0\}$, and a part $\Gamma'' \subset \{x_n > 0\}$. We consider a C^α function, which we denote by φ, which satisfies the boundary conditions

$$\varphi_{|\Gamma''} = 0, \quad \sum_{k=1}^{n} a_{n,k}(x) D_k \varphi(x)_{|\Gamma'} = 0.$$

We want to estimate the integral $\int_{\mathcal{O}'} |D\varphi|^{2p} dx$, $p > 1$. We set

$$\frac{\partial \varphi}{\partial n}(x) = \sum_{k=1}^{n} a_{n,k}(x) D_k \varphi(x).$$

We can first notice that

$$|D\varphi(x)|^{2p} \leq K \left(\sum_{k=1}^{n-1} |D_k \varphi(x)|^{2p} + \left| \frac{\partial \varphi}{\partial n}(x) \right|^{2p} \right)$$

then for $k = 1, \ldots, n-1$ we have

$$\int_{\mathcal{O}'} |D_k \varphi|^{2p} dx = -(2p-1) \int_{\mathcal{O}'} \varphi D_k^2 \varphi |D_k \varphi|^{2p-2} dx.$$

This is thanks to the fact that the integration leads to boundary terms on Γ'', since $k \neq n$. This leads to

$$\int_{\mathcal{O}'} |D_k \varphi|^{2p} dx \leq K_p \int_{\mathcal{O}'} |\varphi D_k^2 \varphi|^p dx.$$

If the domain is small, in view of the fact that φ is C^α and vanishes on Γ'', we can also bound $||\varphi||_{L^\infty}$ by a small number. So we obtain the estimate

$$\int_{\mathcal{O}'} |D_k \varphi|^{2p} dx \leq \varepsilon \int_{\mathcal{O}'} |D_k^2 \varphi|^p dx.$$

Next we write

$$\int_{\mathcal{O}'} \left| \frac{\partial \varphi}{\partial n}(x) \right|^{2p} dx = \int_{\mathcal{O}'} \left| \frac{\partial \varphi}{\partial n}(x) \right|^{2p-1} \left(\sum_{k=1}^{n} a_{n,k}(x) D_k \varphi(x) \right) dx.$$

The terms with index $k \neq n$ can be estimated as above. There remains the term $\int_{O'} \left| \frac{\partial \varphi}{\partial n}(x) \right|^{2p-1} D_n \varphi(x) dx$. The parts integration is still possible, because on the part of the boundary Γ' we have $\frac{\partial \varphi}{\partial n}(x) = 0$. Eventually we get an estimate

$$\int_{O'} |D\varphi|^{2p} dx \leq \varepsilon \int_{O'} |D^2 \varphi|^p dx + \varepsilon.$$

The reason for the addtional term stems from the derivatives of the functions $a_{n,k}(x)$. This generic estimate is used together with local charts and addtional localization to work with small size domains. Collecting results, we can obtain the estimate (10.24) up to the boundary. $\square$

10.3   A Priori Estimates for m

We build on the results obtained in the preceding section. Under the assumptions of Proposition 11, we can assert that $Du \in L^\infty(L^\infty)$. Therefore, the vectors $G^i(x, Du)$ are bounded functions. So we can look at the functions m^i as solving a generic problem stated as follows [we drop the index i and we replace $G^i(x, Du)$ by a bounded vector $G(x,t)$, simply referred as $G(x)$]

$$\frac{\partial m}{\partial t} + A^* m + \operatorname{div}(mG(x)) = 0, \, x \in \mathcal{O}$$

$$\sum_{l=1}^{n} v_l(x) \left(\sum_{k=1}^{n} \frac{\partial}{\partial x_k}(a_{kl}(x)m) - G_l m \right) = 0, \, x \in \partial \mathcal{O}$$

$$m(x,0) = m_0(x) \qquad\qquad (10.27)$$

and m_0 is in $L^2(\mathcal{O})$. We write also (10.27) in the weak form

$$\int_{\mathcal{O}} \frac{\partial m}{\partial t} \varphi dx + \sum_{k,l} \int_{\mathcal{O}} a_{kl} \frac{\partial m}{\partial x_k} \frac{\partial \varphi}{\partial x_l} dx + \sum_l \int_{\mathcal{O}} m \left(\sum_k \frac{\partial a_{kl}}{\partial x_k} - G_l \right) \frac{\partial \varphi}{\partial x_l} dx = 0,$$

$$\forall \varphi \in W^{1,2}. \qquad\qquad (10.28)$$

10.3.1 $L^2(W^{1,2})$ Estimate

We have the following

Proposition 12. *We assume G bounded, and $m_0 \in L^2(\mathcal{O})$. Then we have*

$$||m||_{L^\infty(L^2)} + ||m||_{L^2(W^{1,2})} \leq C, \tag{10.29}$$

where C depends only on the L^∞ bound of G, on the $W^{1,\infty}$ bound of $a_{i,j}$, the ellipticity constant, and of the L^2 norm of m_0.

Proof. This is easily obtained, by taking $\varphi = m(t)$ in (10.28) and integrating in t. The result follows Gronwall's inequality.   □

10.3.2 $L^\infty(L^\infty)$ *Estimates*

First, it is easy to check that, if $m_0 \in L^p(\mathcal{O})$, then

$$||m||_{L^\infty(L^p)} \leq C_p, \forall 2 \leq p < \infty. \tag{10.30}$$

This is obtained by testing (10.28) with m^{p-1}, performing easy bounds, and using Gronwall's inequality [an intermediate step to justify the existence of integrals is done by replacing m by $\min(m, L)$]. However the constant C_p depends on p and goes to ∞ as $p \to \infty$. This is due to the third term in (10.28). In the case when

$$\sum_k \frac{\partial a_{kl}}{\partial x_k} - G_l = 0 \tag{10.31}$$

then we have simply

$$\int_{\mathcal{O}} m^p(x,t)dx \leq \int_{\mathcal{O}} m_0^p(x)dx.$$

Therefore,

$$||m||_{L^\infty(L^p)} \leq ||m_0||_{L^\infty}$$

and we can let $p \to \infty$ and obtain

$$||m||_{L^\infty(L^\infty)} \leq ||m_0||_{L^\infty}. \tag{10.32}$$

The $L^\infty(L^\infty)$ estimate for the general case requires an argument given by Moser (see [4], e.g., for the elliptic case). We provide some details. We set

$$\tilde{G}_l(x) = G_l(x) - \sum_k \frac{\partial a_{kl}}{\partial x_k}(x).$$

We have the following

Proposition 13. *We make the assumptions of Proposition 14, then $m \in L^\infty(L^\infty)$, and the bound depends only on the L^∞ bound of G, the $W^{1,\infty}$ bound of $a_{i,j}$, the ellipticity constant, and the L^∞ norm of m_0.*

Proof. We reduce the initial condition m_0 to be 0, by introducing $\tilde{m}$ to be the solution of

$$\int_{\mathcal{O}} \frac{\partial \tilde{m}}{\partial t} \varphi dx + \sum_{k,l} \int_{\mathcal{O}} a_{kl} \frac{\partial \tilde{m}}{\partial x_k} \frac{\partial \varphi}{\partial x_l} dx = 0$$

$$\tilde{m}(x,0) = m_0(x). \tag{10.33}$$

$\square$

We have seen above that $\tilde{m} \in L^\infty(L^\infty)$. We set $\pi = m - \tilde{m}$. Then π is the solution of

$$\int_{\mathcal{O}} \frac{\partial \pi}{\partial t} \varphi dx + \sum_{k,l} \int_{\mathcal{O}} a_{kl} \frac{\partial \pi}{\partial x_k} \frac{\partial \varphi}{\partial x_l} dx - \sum_l \int_{\mathcal{O}} (\pi + \tilde{m}) \tilde{G}_l \frac{\partial \varphi}{\partial x_l} dx = 0,$$

$$\forall \varphi \in W^{1,2}, \tag{10.34}$$

and $\pi(x,0) = 0$.

We take $p > 2$ and use as a test function in (10.34) $\varphi = |\pi|^{p-2}\pi$. We obtain easily

$$\frac{1}{p} \frac{d}{dt} \int_{\mathcal{O}} |\pi(x,t)|^p dx + (p-1) \sum_{k,l} \int_{\mathcal{O}} a_{kl} \frac{\partial \pi}{\partial x_k} \frac{\partial \pi}{\partial x_l} |\pi|^{p-2} dx - (p-1)$$

$$\sum_k \int_{\mathcal{O}} (\pi + \tilde{m}) \tilde{G}_k \frac{\partial \pi}{\partial x_k} |\pi|^{p-2} dx = 0$$

and we can write

$$\frac{1}{p} \frac{d}{dt} \int_{\mathcal{O}} |\pi(x,t)|^p dx + (p-1)\alpha \int_{\mathcal{O}} |D|\pi||^2 |\pi|^{p-2} dx \leq (p-1)c$$

$$\int_{\mathcal{O}} |\pi|^{p-2}(|\pi|+1)|D|\pi|| dx$$

hence, by standard estimation,

$$\frac{1}{p} \frac{d}{dt} \int_{\mathcal{O}} |\pi(x,t)|^p dx + (p-1)\frac{\alpha}{2} \int_{\mathcal{O}} |D|\pi||^2 |\pi|^{p-2} dx \leq (p-1)\frac{c^2}{\alpha}$$

$$\int_{\mathcal{O}} |\pi|^{p-2}(1 + |\pi|^2) dx.$$

We use

$$\int_{\mathcal{O}} |D|\pi||^2 |\pi|^{p-2} dx = \frac{4}{p^2} \int_{\mathcal{O}} |D|\pi|^{\frac{p}{2}}|^2 dx.$$

We next use Poincaré's inequality to state $(n > 2)$

$$\int_{\mathcal{O}} |D|\pi|^{\frac{p}{2}}|^2 dx \geq c_1 \left(\int_{\mathcal{O}} \left(|\pi|^{\frac{p}{2}} - \int_{\mathcal{O}} |\pi|^{\frac{p}{2}} dx \right)^{\frac{2n}{n-2}} dx \right)^{\frac{n-2}{n}}.$$

If $n = 2$ we replace $\frac{n}{n-2}$ by any $q_0 > 2$.

After easy calculations

$$\int_{\mathcal{O}} |D|\pi|^{\frac{p}{2}}|^2 dx \geq k \left(\int_{\mathcal{O}} |\pi|^{\frac{pn}{n-2}} dx \right)^{\frac{n-2}{n}} - k' \int_{\mathcal{O}} |\pi|^p dx.$$

Collecting results, we obtain the inequality (recall that $p > 2$)

$$\frac{d}{dt} \int_{\mathcal{O}} |\pi(x,t)|^p dx + \left(\int_{\mathcal{O}} |\pi|^{\frac{pn}{n-2}} dx \right)^{\frac{n-2}{n}} \leq \beta p^2 \left[\int_{\mathcal{O}} |\pi|^{p-2} (1 + |\pi|^2) dx \right]. \quad (10.35)$$

We deduce easily (modifying β)

$$\sup_{0 \leq t \leq T} \int_{\mathcal{O}} |\pi(x,t)|^p dx + \int_0^T \left(\int_{\mathcal{O}} |\pi(x,t)|^{\frac{pn}{n-2}} dx \right)^{\frac{n-2}{n}} dt \leq \beta p^2$$

$$\left[\int_0^T \int_{\mathcal{O}} |\pi(x,t)|^{p-2} (1 + |\pi|^2) dx dt \right]. \quad (10.36)$$

Multiplying by $(\sup_{0 \leq t \leq T} \int_{\mathcal{O}} |\pi(x,t)|^p dx)^{\frac{2}{n}}$ we obtain

$$\left(\sup_{0 \leq t \leq T} \int_{\mathcal{O}} |\pi(x,t)|^p dx \right)^{1+\frac{2}{n}} + \left(\sup_{0 \leq t \leq T} \int_{\mathcal{O}} |\pi(x,t)|^p dx \right)^{\frac{2}{n}}$$

$$\int_0^T \left(\int_{\mathcal{O}} |\pi(x,t)|^{\frac{pn}{n-2}} dx \right)^{\frac{n-2}{n}} dt \leq \left(\sup_{0 \leq t \leq T} \int_{\mathcal{O}} |\pi(x,t)|^p dx \right)^{\frac{2}{n}} \beta p^2$$

$$\left[\int_0^T \int_{\mathcal{O}} |\pi(x,t)|^{p-2} (1 + |\pi|^2) dx dt \right].$$

We next use the inequality

$$\int_0^T \int_{\mathcal{O}} |p(x,t)|^{p(1+\frac{2}{n})} dx d\theta \leq \left(\sup_{0 \leq t \leq T} \int_{\mathcal{O}} |p(x,t)|^p dx \right)^{\frac{2}{n}} \int_0^T \left(\int_{\mathcal{O}} |\pi(x,t)|^{\frac{pn}{n-2}} dx \right)^{\frac{n-2}{n}} dt$$

therefore, we obtain,

$$\int_0^T \int_{\mathcal{O}} |\pi(x,t)|^{p(1+\frac{2}{n})} dxdt \leq (\beta p^2)^{1+\frac{2}{n}} \left[\int_0^T \int_{\mathcal{O}} |\pi(x,t)|^{p-2}(1+|\pi|^2)dxdt \right]^{1+\frac{2}{n}}.$$

$$(10.37)$$

We can write this relation as

$$||\pi||_{L^{p(1+\frac{2}{n})}(\mathcal{O}\times(0,T))} \leq \beta^{\frac{1}{p}} p^{\frac{2}{p}} \left[\int_0^T \int_{\mathcal{O}} |\pi(x,t)|^{p-2}(1+|\pi|^2)dxdt \right]^{\frac{1}{p}}. \quad (10.38)$$

We use

$$\int_0^T \int_{\mathcal{O}} |\pi(x,t)|^{p-2} dxdt \leq \left(\int_0^T \int_{\mathcal{O}} |\pi(x,t)|^p dxdt \right)^{\frac{p-2}{p}} (T|\mathcal{O}|)^{\frac{2}{p}}$$

$$\leq \max \left(1, \int_0^T \int_{\mathcal{O}} |\pi(x,t)|^p dxdt \right) (T|\mathcal{O}|)^{\frac{2}{p}}.$$

Therefore,

$$\int_0^T \int_{\mathcal{O}} |\pi(x,t)|^{p-2}(1+|\pi|^2)dxdt$$

$$\leq \max \left(1, \int_0^T \int_{\mathcal{O}} |\pi(x,t)|^p dxdt \right) (1+(T|\mathcal{O}|)^{\frac{2}{p}})$$

$$\leq \max \left(1, \int_0^T \int_{\mathcal{O}} |\pi(x,t)|^p dxdt \right) (1+\max(1+T|\mathcal{O}|))$$

hence also

$$\left[\int_0^T \int_{\mathcal{O}} |\pi(x,t)|^{p-2}(1+|\pi|^2)dxdt \right]^{\frac{1}{p}} \leq \max(1, ||\pi||_{L^p}) c^{\frac{1}{p}}.$$

Using this estimate in (10.38) and modifying the constant β we can write

$$||\pi||_{L^{p(1+\frac{2}{n})}} \leq \beta^{\frac{1}{p}} p^{\frac{2}{p}} \max(1, ||\pi||_{L^p}).$$

Since 1 is smaller than the right-hand side, we deduce

$$\max(1, ||\pi||_{L^{p(1+\frac{2}{n})}}) \leq \beta^{\frac{1}{p}} p^{\frac{2}{p}} \max(1, ||\pi||_{L^p}). \quad (10.39)$$

This is a classical inequality, which easily leads to the result. Indeed, set $a = 1 + \frac{2}{n}$ and $p_j = 2a^j$, $z_j = \max(1, ||\pi||_{L^{p_j}})$. We can write

$$z_{j+1} \leq \beta^{\frac{1}{p_j}} p_j^{\frac{2}{p_j}} z_j$$

and

$$z_j \leq z_0 \beta^{\Sigma_{h=0}^{\infty} \frac{1}{p_h}} \exp\left(\sum_{h=0}^{\infty} \frac{2\log p_h}{p_h}\right).$$

Since $a > 1$, the series are converging. Therefore, z_j is bounded. Letting $j \to \infty$, we obtain that $\|\pi\|$ is finite. This concludes the proof. $\qquad\square$

10.3.3 Further Estimates

With the $L^{\infty}(L^{\infty})$ estimate, we can see that

$$\sum_l \int_{\mathcal{O}} m\left(\sum_k \frac{\partial a_{kl}}{\partial x_k} - G_l\right) \frac{\partial \varphi}{\partial x_l} dx$$

can be extended to φ in $W^{1,p'}, \forall 1 < p' < 2$. We can consider this term as a right-hand side for (10.28), belonging to $L^{\infty}((W^{1,p'})^*)$.

With the methods of Sect. 10.2.3, we can obtain that m is C^{α} up to the boundary. We can then obtain that $m \in L^p(W_{\text{loc}}^{1,p}), \forall 2 \leq p < \infty$. Although, we believe that $m \in L^p(W^{1,p})$, we do not have a reference in the literature related to the regularity theory of parabolic equations with Neumann boundary conditions to assert the regularity up to the boundary.

So we state the following:

Proposition 14. *We assume G bounded, and $m_0 \in L^{\infty}(\mathcal{O})$. Then $m \in C^{\alpha} \cap L^p(W_{\text{loc}}^{1,p})$, $\dot{m} \in L^p((W_{\text{loc}}^{1,p})^*)$, $\forall 2 \leq p < \infty$. The norms in these functional spaces depends only on the L^{∞} bound of G, on the $W^{1,\infty}$ bound of $a_{i,j}$, the ellipticity constant, and of the L^{∞} norm of m_0.*

10.3.4 Statement of the Global A Priori Estimate Result

We can collect all previous results in a theorem that synthesizes the a priori estimates for the system (10.1) and (10.2).

Theorem 15. *We make the assumptions of Proposition 11, and $m_0^i \in L^{\infty}$; then a weak solution of the system (10.13) and (10.14) in the functional space (10.12) satisfies the regularity properties*

$$u \in L^p(W^{2,p}), \dot{u} \in L^p(L^p)$$

$$m \in C^{\alpha} \cap L^p(W^{1,p}_{\text{loc}}), \dot{m} \in L^p((W^{1,p'}_{\text{loc}})^*)$$

$$\forall 2 \le p < +\infty. \tag{10.40}$$

The norms in the various functional spaces depend only on the $L^{\infty}(L^1)$ norm of m, the L^{∞} norm of m_0, of the various constants, and of the domain.

Corollary 16. *Under the assumptions of Proposition 11, and $m_0^i \in L^{\infty}$, $m_0^i(x) \ge 0$, $\int_{\mathcal{O}} m_0^i(x)dx = 1$, then a weak solution of the system (10.13) and (10.14) in the functional space (10.12), such that $m^i(x,t) \ge 0$, satisfies the regularity properties*

$$u \in L^p(W^{2,p}), \dot{u} \in L^p(L^p)$$

$$m \in C^{\alpha} \cap L^p(W^{1,p}_{\text{loc}}), \dot{m} \in L^p((W^{1,p'}_{\text{loc}})^*)$$

$$\forall 2 \le p < +\infty \tag{10.41}$$

The norms in the various functional spaces depend only on the L^{∞} norm of m_0, the $W^{1,\infty}$ bound of $a_{i,j}$, the ellipticity constant, the number $k_0(1)$, and the domain.

Proof. We simply note that taking

$$\varphi^i(x,t) = \chi^i(t)$$

in (10.14) we can write

$$-\int_0^T \dot{\chi}^i(t) \left(\int_{\mathcal{O}} m^i(x,t)dx \right) dt = \chi^i(0)$$

from which it follows easily that

$$\int_{\mathcal{O}} m^i(x,t)dx = 1, \forall t.$$

Since $m^i(x,t) \ge 0$, $m \in L^{\infty}(L^1)$, and the theorem applies.   □

10.4   Existence Result

We want to prove the following:

Theorem 17. *Under the assumptions of Corollary 16, there exists a solution (u,m) of (10.12)–(10.14), such that $m^i(x,t) \ge 0$ and that satisfies the regularity properties (10.41).*

The proof will consist of two parts. In the first part we consider an approximation of the problem as follows. Define

$$H_\varepsilon^i(x,q) = \frac{H^i(x,q)}{1+\varepsilon|q|^2}, \; G_\varepsilon^i(x,q) = \frac{G^i(x,q)}{1+\varepsilon|q|}$$

$$\beta_\varepsilon(m)(x) = \frac{m(x)}{1+\varepsilon|m(x)|}, \; \beta_\varepsilon^+(m)(x) = \frac{m^+(x)}{1+\varepsilon|m(x)|}$$

$$f_{0,\varepsilon}^i(x,m) = f_0^i(x,\beta_\varepsilon(m)), \; h_\varepsilon^i(x,m) = h^i(x,\beta_\varepsilon(m)).$$

Note that in $\beta_\varepsilon(m)$, m can be a vector in $L^p(\mathcal{O},R^N)$, whereas we shall apply $\beta_\varepsilon^+(m)$ only to a function in $L^p(\mathcal{O})$. We consider the following approximate problem

$$u_\varepsilon, m_\varepsilon \in L^2(W^{1,2}) \cap L^\infty(L^2)$$

$$m_\varepsilon \in C(L^2) \tag{10.42}$$

$$\int_0^T (\dot{u}_\varepsilon^i, \varphi^i)_{L^2}dt + \sum_{k,l=1}^n \int_0^T (D_l u_\varepsilon^i, D_k(a_{kl}\varphi^i))_{L^2}dt = (h_\varepsilon^i(.,m_\varepsilon(T)), \varphi^i(T))_{L^2}$$

$$+ \int_0^T (H_\varepsilon^i(.,Du_\varepsilon) + f_{0\varepsilon}^i(.,m_\varepsilon(t)), \varphi^i)_{L^2}dt, \; \forall i = 1,\ldots,N \tag{10.43}$$

$$-\int_0^T (\dot{m}_\varepsilon^i, \varphi^i)_{L^2}dt + \sum_{k,l=1}^n \int_0^T (D_l\varphi^i, D_k(a_{kl}m_\varepsilon^i))_{L^2}dt$$

$$= (m_0, \varphi^i(0))_{L^2} + \int_0^T (\beta_\varepsilon^+(m_\varepsilon^i)G_\varepsilon^i(.,Du_\varepsilon), D\varphi^i)_{L^2}dt, \; \forall i = 1,\ldots,N \tag{10.44}$$

Proposition 18. *We make the assumptions of Corollary 16. Suppose we have a solution of (10.42)–(10.44); then $m_\varepsilon^i(x,t) \geq 0$ and the pair $u_\varepsilon, m_\varepsilon$ remains bounded in the functional spaces (10.41). We can extract a subsequence, still denoted $u_\varepsilon, m_\varepsilon$, which converges weakly to u,m and also*

$$u_\varepsilon, Du_\varepsilon \to u, Du \text{ a.e. } x,t$$

$$m_\varepsilon \to m \text{ a.e. } x,t$$

$$u_\varepsilon(.,T) \to u(.,T)$$

$$m_\varepsilon(.,T) \to m(.,T), \text{ in } L^2$$

and u,m is a solution of (10.12)–(10.14).

We first note that, since the derivatives $\dot{u}^i_\varepsilon, \dot{m}^i_\varepsilon$ are defined in $L^p(L^p)$ and $L^p((W^{1,p'})^*)$, respectively, we can write the weak forms (10.43) and (10.44) as follows

$$(-\dot{u}^i_\varepsilon, \varphi^i)_{L^2} + \sum_{k,l=1}^n (D_l u^i_\varepsilon, D_k(a_{kl}\varphi^i))_{L^2} = (H^i_\varepsilon(.,Du_\varepsilon) + f^i_{0\varepsilon}(.,m_\varepsilon(t)), \varphi^i)_{L^2} \quad (10.45)$$

$$\forall \varphi^i \in W^{1,2}, \forall i \in 1,\ldots,N$$

$$u^i_\varepsilon(x,T) = h^i_\varepsilon(.,m_\varepsilon(T))$$

and

$$(\dot{m}^i_\varepsilon, \varphi^i)_{L^2} + \sum_{k,l=1}^n (D_l\varphi^i, D_k(a_{kl}m^i_\varepsilon))_{L^2} = (\beta^+_\varepsilon(m^i_\varepsilon)G^i_\varepsilon(.,Du_\varepsilon), D\varphi^i)_{L^2} \quad (10.46)$$

$$\forall \varphi^i \in W^{1,2} \ \forall i = 1,\ldots,N$$

$$m^i_\varepsilon(x,0) = m_0(x).$$

We then prove the positivity property. We take $\varphi^i = (m^i_\varepsilon)^-$ in (10.46). The right-hand side vanishes, and we obtain

$$\frac{1}{2}\frac{d}{dt}|(m^i_\varepsilon)^-|^2 + \sum_{k,l=1}^n (D_l(m^i_\varepsilon)^-, D_k(a_{kl}(m^i_\varepsilon)^-))_{L^2} = 0$$

and $(m^i_\varepsilon)^-(x,0) = 0$. It is easy to check that $(m^i_\varepsilon)^-(.,t) = 0$, $\forall t$. But testing with $\varphi^i = 1$, we get

$$\int_{\mathcal{O}} m^i_\varepsilon(x,t)dx = \int_{\mathcal{O}} m_0(x)dx = 1$$

and thus

$$\|f^i_{0\varepsilon}(.,m_\varepsilon(t))\|_{L^\infty} \le k_0(1)$$

$$\|h^i_\varepsilon(.,m_\varepsilon(T))\|_{L^\infty} \le k_0(1).$$

Since $H_\varepsilon^i(x,q)$ satisfies the same assumptions as $H^i(x,q)$ in terms of growth, uniformly in ε, we obtain

$$||u_\varepsilon||_{L^\infty(L^\infty)}, ||Du_\varepsilon||_{L^p(W^{1,p})}, ||\dot{u}_\varepsilon||_{L^p(L^p)} \leq C$$

hence also

$$||u_\varepsilon||_{C^\delta(\bar{\mathcal{O}}\times[0,T])} \leq C$$

Therefore, for a subsequence

$$u_\varepsilon \rightharpoonup u, \text{ weakly in } L^P(W^{2,p}), \dot{u}_\varepsilon \rightharpoonup u, \text{ weakly in } L^P(L^p)$$

$$u_\varepsilon \to u, \text{ in } C^0(\bar{\mathcal{O}} \times [0,T]).$$

Since $H_\varepsilon^i(.,Du_\varepsilon) + f_{0\varepsilon}^i(.,m_\varepsilon(t))$ is bounded in $L^1(\mathcal{O} \times (0,T))$ we obtain from (10.45)

$$\sum_{k,l=1}^{n} \int_0^T (D_l u_\varepsilon^i, a_{kl} D_k(u_\varepsilon^i - u^i))_{L^2} dt \to 0$$

from which we obtain

$$u_\varepsilon \to u, \text{ in } L^2(W^{1,2}).$$

We can extract a subsequence such that

$$Du_\varepsilon \to Du, \text{ a.e.} x,t$$

and we know that

$$||Du_\varepsilon||_{L^\infty(L^\infty)} \leq C.$$

We next consider the equations for m_ε^i. We can write (10.46) as

$$(\dot{m}_\varepsilon^i, \varphi^i)_{L^2} + \sum_{k,l=1}^{n} (D_l \varphi^i, a_{kl} D_k m_\varepsilon^i)_{L^2} = \sum_{l=1}^{n} (m_\varepsilon^i \tilde{G}_{\varepsilon,l}^i, D_l \varphi^i)_{L^2} \qquad (10.47)$$

with

$$\tilde{G}_{\varepsilon,l}^i(x) = \frac{G_\varepsilon^i(x, Du_\varepsilon(x))}{1 + \varepsilon m_\varepsilon^i(x)} - \sum_k D_k a_{k,l}(x)$$

and the functions $\tilde{G}_{\varepsilon,l}^i$ are bounded in L^∞, uniformly in ε. Therefore,

$$||m_\varepsilon||_{L^P(W^{1,p})} \leq C, ||\dot{m}_\varepsilon||_{L^P((W^{1,p})^*)} \leq C$$

$$||m_\varepsilon||_{L^\infty(L^\infty)} \leq C.$$

We deduce that for a subsequence

$$m_\varepsilon \to m, \text{ a.e. } x, t$$

and

$$m_\varepsilon^i \tilde{G}_{\varepsilon,l}^i(x) \to m^i(x) G_l^i(x, Du), \text{ a.e. } x, t.$$

From (10.47) and with an argument similar to that used for u, we see that

$$m_\varepsilon \to m, \text{ in } L^2(W^{1,2})$$

$$m_\varepsilon(.,t) \to m(.,t), \text{ in } L^2, \forall t.$$

We then obtain that

$$h_\varepsilon^i(., m_\varepsilon(T)) \to h^i(., m(T)), \text{ in } L^2.$$

Collecting results we see that u, m is a solution of (10.12)–(10.14) and $m^i(x,t) \geq 0$.
□

To complete the proof of Theorem 17, it remains to prove the existence of a solution of (10.42), (10.45), and (10.46). Now ε is fixed. We omit to mention it for the unknown functions $u_\varepsilon, m_\varepsilon$. We use the following notation

$$\Phi^i(u,m)(x,t) = H_\varepsilon^i(x, Du) + f_{0\varepsilon}^i(x,m)$$

$$\Psi^i(u,m)(x,t) = \beta_\varepsilon^+(m^i) G_\varepsilon^i(., Du)$$

and call $h^i(x,m)$ the functional $h_\varepsilon^i(x,m)$.

The functionals Φ^i, Ψ^i map $W^{1,2}(\mathcal{O}, \mathbb{R}^N) \times L^2(\mathcal{O}, \mathbb{R}^N)$ into $L^\infty(\mathcal{O})$ and $L^\infty(\mathcal{O}, \mathbb{R}^n)$, respectively (they may depend on time). The system (10.45) and (10.46) becomes

$$(-\dot{u}^i, \varphi^i)_{L^2} + \sum_{k,l=1}^n (D_l u^i, D_k(a_{kl}\varphi^i))_{L^2} = (\Phi^i(u,m), \varphi^i)_{L^2}$$

$$u^i(x,T) = h^i(x, m(T)) \tag{10.48}$$

$$(\dot{m}^i, \varphi^i)_{L^2} + \sum_{k,l=1}^{n} (D_l \varphi^i, a_{kl} D_k m^i)_{L^2} = \sum_{l=1}^{n} (\Psi_l^i(u,m), D_l \varphi^i)_{L^2}$$

$$m^i(x,0) = m_0(x). \tag{10.49}$$

To solve this system, we use the Galerkin approximation. We consider an orthonormal basis of $L^2(\mathcal{O})$, made of functions of $W^{1,2}(\mathcal{O})$, namely $\varphi_1, \ldots, \varphi_r, \ldots$. We approximate u^i, m^i by $u^{i,R}, m^{i,R}$ defined by

$$u^{i,R}(x,t) = \sum_{r=1}^{R} c_r^i(t) \varphi_r(x)$$

$$m^{i,R}(x,t) = \sum_{r=1}^{R} b_r^i(t) \varphi_r(x).$$

The new unknowns are the functions $c_r^i(t), b_r^i(t)$. They will be solutions of a system of forward, backward differential equations

$$-\frac{dc_r^i}{dt} + \sum_{\rho=1}^{R} \sum_{k,l=1}^{n} (D_k(a_{k,l}\varphi_r), D_l \varphi_\rho)_{L^2} c_\rho^i = (\Phi^i(u^R, m^R), \varphi_r)_{L^2}$$

$$c_r^i(T) = (h^i(., m^R(T)), \varphi_r)_{L^2} \tag{10.50}$$

$$\frac{db_r^i}{dt} + \sum_{\rho=1}^{R} \sum_{k,l=1}^{n} (D_l \varphi_r, a_{k,l} D_k \varphi_\rho)_{L^2} b_\rho^i = \sum_{l=1}^{n} (\Psi_l^i(u^R, m^R), D_l \varphi_r)_{L^2}$$

$$b_r^i(0) = (m_0, \varphi_r)_{L^2}. \tag{10.51}$$

We begin by checking a priori estimates. We recall that $\Phi^i(u^R, m^R)$ and $\Psi_l^i(u^R, m^R)$ are bounded by an absolute constant. We multiply (10.50) by c_r^i and sum up over r. We get

$$-\frac{1}{2}\frac{d}{dt}|u^{i,R}|^2 + \sum_{k,l=1}^{n} (D_k(a_{k,l}u^{i,R}), D_l u^{i,R})_{L^2} = (\Phi^i(u^R, m^R), u^{i,R})_{L^2}$$

$$u^{i,R}(T) = \sum_{r=1}^{R} (h^i(., m^R(T)), \varphi_r)_{L^2} \varphi_r$$

from which we deduce easily that

$$\|u^R\|_{L^2(W^{1,2})} + \|u^R\|_{L^\infty(L^2)} \leq C. \tag{10.52}$$

Similarly, from (10.51) we can write

$$\frac{1}{2}\frac{d}{dt}|m^{i,R}|^2 + \sum_{k,l=1}^{n}(a_{k,l}D_k m^{i,R}, D_l m^{i,R})_{L^2} = \sum_{l=1}^{n}(\Psi_l^i(u^R, m^R), D_l m^{i,R})_{L^2}$$

$$m^{i,R}(0) = \sum_{r=1}^{R}(m_0, \varphi_r)_{L^2}\varphi_r$$

and also

$$||m^R||_{L^2(W^{1,2})} + ||m^R||_{L^\infty(L^2)} \leq C. \tag{10.53}$$

It follows in particular, from these estimates that the functions $c_r^i(t), b_r^i(t)$ are bounded by an absolute constant. From the differential equations we see that the derivatives are also bounded, however, by a constant, which this time depends on R. These a priori estimates are sufficient to prove that the system (10.50) and (10.51) has a solution. Indeed, one uses a fixed-point argument. Here R is fixed. Given u^R, m^R on the right-hand side of (10.50) and (10.51), we can solve (10.50) and (10.51) independently. They are linear differential equations. One is forward, the other one is backward, but they are uncoupled. In this way, one defines a map from a compact susbet of $(C[0,T])^R \times (C[0,T])^R$ into itself. The compactness is provided by the Arzela–Ascoli theorem. The map is continuous. So it has a fixed point, which is a solution of (10.50) and (10.51). The estimates (10.52) and (10.53) hold.

We also have the estimates

$$\frac{1}{h}\int_0^{T-h}\int_{\mathcal{O}}|u^{i,R}(x,t+h) - u^{i,R}(x,t)|^2 dx dt \leq C$$

$$\frac{1}{h}\int_0^{T-h}\int_{\mathcal{O}}|m^{i,R}(x,t+h) - m^{i,R}(x,t)|^2 dx dt. \tag{10.54}$$

Proof of (10.54) This is a classical property. We check only the first one. We first write

$$-\frac{d}{dt}u^{i,R} + \sum_{r=1}^{R}\sum_{k,l=1}^{n}(a_{k,l}D_k\varphi_r, D_l u^{i,R})_{L^2}\varphi_r \tag{10.55}$$

$$= \sum_{r=1}^{R}(\Phi^i(u^R, m^R) - \sum_{k,l=1}^{n}D_k a_{k,l}D_l u^{i,R}, \varphi_r)\varphi_r = g^{i,R}$$

and g^R is bounded in $L^2(L^2)$. It follows that

$$-\frac{1}{h}(u^{i,R}(t+h) - u^{i,R}(t), u^{i,R}(t))_{L^2} + \sum_{k,l=1}^{n}\left(a_{k,l}D_k u^{i,R}(t), D_l\frac{1}{h}\int_t^{t+h}u^{i,R}(s)ds\right)_{L^2}$$

$$= \frac{1}{h} \left(\int_t^{t+h} g^R(s)ds, u^{i,R}(t) \right)_{L^2}.$$

From previous estimates, we obtain

$$\frac{1}{h} \int_t^{T-h} (u^{i,R}(t), u^{i,R}(t) - u^{i,R}(t+h))_{L^2} dt \le C$$

from which the estimate (10.54) follows easily.   □

To proceed, we need to work with a particular basis of $L^2(\mathcal{O})$. We take φ_r to be the eigenvectors of the operator $-\Delta + I$ on the domain $\mathcal{O}$ with Neumann boundary conditions. So φ_r satisfies

$$((\varphi_r, v))_{W^{1,2}(\mathcal{O})} = \lambda_r (\varphi_r, v)_{L^2(\mathcal{O})}, \forall v \in W^{1,2}(\mathcal{O}),$$

where λ_r is the eigenvalue corresponding to φ_r. Define now A^R over $W^{1,2}(\mathcal{O})$ by writing

$$A^R z = \sum_{r=1}^{R} \sum_{k,l=1}^{n} (a_{k,l} D_k \varphi_r, D_l z)_{L^2} \varphi_r$$

then A^R maps $W^{1,2}(\mathcal{O})$ into the subspace of $L^2(\mathcal{O})$ generated by $\varphi_r, r = 1, \dots, R$, with the property

$$\|A^R\|_{\mathcal{L}(W^{1,2}; (W^{1,2})^*)} \le C.$$

This is because $\frac{\varphi_r}{\sqrt{\lambda_r}}$ is an orthonormal basis of $W^{1,2}(\mathcal{O})$ and

$$\left\| \sum_{r=1}^{R} (\varphi_r, v) \varphi_r \right\|_{W^{1,2}} \le \|v\|_{W^{1,2}}.$$

The differential equation (10.55) writes

$$-\frac{d}{dt} u^{i,R} + A^R u^{i,R} = g^{i,R} \tag{10.56}$$

and thus we get

$$\left\| \frac{d}{dt} u^{i,R} \right\|_{L^2((W^{1,2})^*)} \le C, \quad \left\| \frac{d}{dt} m^{i,R} \right\|_{L^2((W^{1,2})^*)} \le C \tag{10.57}$$

the second property being proved similarly. We have also

$$\|u^{i,R}(T)\|_{W^{1,2}} \le C. \tag{10.58}$$

Thanks to the estimates (10.52), (10.54), and (10.58), we can extract a sequence such that

$$u^{i,R} \to u^i \text{ in } L^2(L^2)$$

$$u^{i,R} \rightharpoonup u^i \text{ in } L^2(W^{1,2}) \text{ weakly}$$

$$\frac{d}{dt}u^{i,R} \rightharpoonup \frac{d}{dt}u^i \text{ in } L^2((W^{1,2})^*) \text{ weakly}$$

$$u^{i,R}(T) \to u^i(T) \text{ in } L^2. \tag{10.59}$$

Testing (10.56) with $u^{i,R} - u^i$, we see easily that

$$u^{i,R} \to u^i \text{ in } L^2(W^{1,2}). \tag{10.60}$$

From (10.53) and (10.54) we can assert that, for a subsequence,

$$m^{i,R} \to m^i \text{ in } L^2(L^2). \tag{10.61}$$

Therefore, for a subsequence,

$$Du^{i,R} \to Du^i$$

$$m^{i,R} \to m^i, \text{ a.e.}$$

and then

$$\Phi^i(u^R, m^R) \to \Phi^i(u, m), \text{ a.e.}$$

$$\Psi^i(u^R, m^R) \to \Psi^i(u, m), \text{ a.e.}$$

By a reasoning similar to that done for $u^{i,R}$ we can prove that

$$m^{i,R} \to m^i \text{ in } L^2(W^{1,2}). \tag{10.62}$$

With these convergence results, one can pass to the limit in the Galerkin approximation and check that u, m is a solution of (10.45) and (10.46).   □

References

1. Andersson, D., Djehiche, B. (2011). A maximum principle for SDEs of mean field type. *Applied Mathematics and Optimization, 63,* 341–356.
2. Bardi, M. (2011). Explicit solutions of some linear-quadratic mean field games. *Networks and Heterogeneous Media,* 7(2), 243–261 (2012)
3. Bensoussan, A., Frehse, J. (1990). C^{α} regularity results for quasi-linear parabolic systems. *Commentationes Mathematicae Universitatis Carolinae, 31,* 453–474.
4. Bensoussan, A., Frehse, J. (1995). Ergodic Bellman systems for stochastic games in arbitrary dimension. *Proceedings of the Royal Society, London, Mathematical and Physical sciences A, 449,* 65–67.
5. Bensoussan, A., Frehse, J. (2002). *Regularity results for nonlinear elliptic systems and applications.* Applied mathematical sciences, vol. 151. Springer, Berlin.
6. Bensoussan, A., Frehse, J. (2002). Smooth solutions of systems of quasilinear parabolic equations. *ESAIM: Control, Optimization and Calculus of Variations, 8,* 169–193.
7. Bensoussan, A., Frehse, J. (2013). Control and Nash games with mean field effect. *Chinese Annals of Mathematics, 34B,* 161–192.
8. Bensoussan, A., Frehse, J., Vogelgesang, J. (2010). Systems of Bellman equations to stochastic differential games with noncompact coupling, discrete and continuous dynamical systems. *Series A, 274,* 1375–1390.
9. Bensoussan, A., Frehse, J., Vogelgesang, J. (2012). Nash and stackelberg differential games. *Chinese Annals of Mathematics, Series B, 33*(3), 317–332.
10. Bensoussan, A., Sung, K.C.J., Yam, S.C.P., Yung, S.P. (2011). Linear-quadratic mean field games. Technical report.
11. Björk, T., Murgoci, A. (2010). A general theory of Markovian time inconsistent stochastic control problems. *SSRN: 1694759.*
12. Buckdahn, R., Djehiche, B., Li, J. (2011). A general stochastic maximum principle for SDEs of mean-field type. *Applied Mathematics and Optimization, 64,* 197–216.
13. Buckdahn, R., Li, J., Peng, SG. (2009). Mean-field backward stochastic differential equations and related partial differential equations. *Stochastic Processes and their Applications, 119,* 3133–3154.
14. Cardaliaguet, P. (2010). Notes on mean field games. Technical report.
15. Carmona, R., Delarue, F. (2012). Probabilistic analysis of mean-field games. http://arxiv.org/abs/1210.5780.
16. Garnier, J., Papanicolaou, G., Yang, T. W. (2013). Large deviations for a mean field model of systemic risk. http://arxiv.org/abs/1204.3536.

A. Bensoussan et al., *Mean Field Games and Mean Field Type Control Theory*, SpringerBriefs in Mathematics, DOI 10.1007/978-1-4614-8508-7,
© Alain Bensoussan, Jens Frehse, Phillip Yam 2013

17. Guéant, O., Lasry, J.-M., Lions, P.-L. (2011). Mean field games and applications. In: A.R. Carmona, et al. (Eds.), *Paris-Princeton Lectures on Mathematical Sciences 2010* (pp. 205–266).

18. Hu, Y., Jin, H. Q., Zhou, X. Y. (2011). Time-inconsistent stochastic linear-quadratic control. http://arxiv.org/abs/1111.0818.

19. Huang, M. (2010). Large-population LQG games involving a major player: the nash certainty equivalence principle. *SIAM Journal on Control and Optimization, 48*(5), 3318–3353.

20. Huang, M., Malhamé, R. P., Caines, P. E. (2006). Large population stochastic dynamic games: closed-loop MCKean-Vlasov systems and the nash certainty equivalence principle. *Communications in Information and Systems, 6*(3), 221–252.

21. Huang, M., Caines, P. E., Malhamé, R. P. (2007). Large-population cost-coupled LQG problems with nonuniform agents: individual-mass behavior and decentralized ε−nash equilibria. *IEEE Transactions on Automatic Control, 52*(9), 1560–1571.

22. Huang, M., Caines, P. E., Malhamé, R. P. (2007). An invariance principle in large population stochastic dynamic games. *Journal of Systems Science and Complexity, 20*(2), 162–172.

23. Kolokoltsov, V. N., Troeva, M., Yang, W. (2012). On the rate of convergence for the mean-field approximation of controlled diffusions with a large number of players. Working paper.

24. Kolokoltsov, V. N., Yang, W. (2013). Sensitivity analysis for HJB equations with an application to a coupled backward-forward system. Working paper.

25. Lasry, J.-M., Lions, P.-L. (2006). Jeux à champ moyen I- Le cas stationnaire. *Comptes Rendus de l'Académie des Sciences, Series I, 343*, 619–625.

26. Lasry, J.-M., Lions, P.-L. (2006). Jeux à champ moyen II- Horizn fini et contrôle optimal, *Comptes Rendus de l'Académie des Sciences, Series I, 343*, 679–684.

27. Lasry, J.-M., Lions, P.-L. (2007). Mean field games. *Japanese Journal of Mathematics, 2*(1), 229–260.

28. Ladyzhenskaya , O., Solonnikov, V., Uraltseva, N. (1968). *Linear and Quasi-linear Equations of Parabolic Type*. In: Translations of Mathematical Monographs, vol. 23. American Mathematical Society, Providence.

29. McKean, H. P. (1966). A class of Markov processes associated with nonlinear parabolic equations. *Proceedings of the National Academy of Sciences USA, 56*, 1907–1911.

30. Meyer-Brandis, T., Øksendal, B., Zhou, X. Z. (2012). A mean-field stochastic maximum principle via Malliavin calculus. *Stochastics (A Special Issue for Mark Davis' Festschrift), 84*, 643–666.

31. Nourian, M., Caines, P. E. (2012). ε-nash mean field game theory for nonlinear stochastic dynamical systems with major and minor agents. *SIAM Journal on Control and Optimization* (submitted).

32. Peng, S. (1992). Stochastic Hamilton-Jacobi-Bellman equations. *SIAM Journal on Control and Optimization, 30*(2), 284–304.

33. Schlag, W. (1994). Schauder and L^p−estimates for Parabolic systems via Campanato spaces. *Comm. P.D.E., 21(7–8)*, 1141–1175.

Index

A
A priori estimates, 4, 104–118, 123, 124
Approximate Nash games, 31–43

B
Backward stochastic differential equations, 14

C
Campanato test, 107
Coalitions, 4, 91
Community (large), 1, 4, 91, 97–99
Compatibility condition, 109
Cost functional, 60, 71, 72

D
Differential games, 4, 34, 38, 43, 57, 59, 70, 71, 73, 91–99
Dirac measure, 7, 31
Dirichlet boundary condition, 102, 109
Dirichlet problem, 106
Drift, 83
Dual control problems, 3, 91
Dynamic programming, 2, 13, 18, 25, 33, 71

E
Elliptic operator, 106, 108
Elliptic systems, 3
Ergodic controls, 3

F
Feedback controls, 7, 18, 32, 37, 51, 60, 65, 91
Fixed point, 2, 4, 61, 122

Fokker–Planck (FP) equations, 2, 3, 11
Forward-backward stochastic differential equations, 14
Frechet derivative, 52, 63

G
Galerkin approximation, 123, 126
Gateaux derivative, 18, 19
Gateaux differential, 18, 61, 77, 79, 82, 93
Green's function, 105, 106

H
Hamilton–Jacobi–Bellman (HJB) equation, 2–4, 11–13, 25, 46, 49, 56, 62, 70, 83, 95
Hamiltonian (function), 11, 13, 22, 34, 59, 62, 70, 92, 105, 106
Hölder estimates, 107

I
Initial condition, 3, 25, 26, 28, 55–57, 114
Ito's formula, 13, 27, 32

L
Lagrangian (function), 11
Law of large numbers, 36, 66, 73
Lebesgue's theorem, 36, 98
Linear operator, 12
Linear-quadratic, 2, 3, 45–57, 84
Lp estimates, 108–112

A. Bensoussan et al., *Mean Field Games and Mean Field Type Control Theory*,
SpringerBriefs in Mathematics, DOI 10.1007/978-1-4614-8508-7,
© Alain Bensoussan, Jens Frehse, Phillip Yam 2013